Progress in IS

"PROGRESS in IS" encompasses the various areas of Information Systems in theory and practice, presenting cutting-edge advances in the field. It is aimed especially at researchers, doctoral students, and advanced practitioners. The series features both research monographs that make substantial contributions to our state of knowledge and handbooks and other edited volumes, in which a team of experts is organized by one or more leading authorities to write individual chapters on various aspects of the topic. "PROGRESS in IS" is edited by a global team of leading IS experts. The editorial board expressly welcomes new members to this group. Individual volumes in this series are supported by a minimum of two members of the editorial board, and a code of conduct mandatory for all members of the board ensures the quality and cutting-edge nature of the titles published under this series.

Melissa Lukings • Arash Habibi Lashkari

Understanding Cybersecurity Law in Data Sovereignty and Digital Governance

An Overview from a Legal Perspective

Springer

Melissa Lukings
York University
Toronto, ON, Canada

Arash Habibi Lashkari
York University
Toronto, ON, Canada

ISSN 2196-8705 ISSN 2196-8713 (electronic)
Progress in IS
ISBN 978-3-031-14263-5 ISBN 978-3-031-14264-2 (eBook)
https://doi.org/10.1007/978-3-031-14264-2

This Springer imprint is published by the registered company Springer Nature Switzerland AG
The registered company address is: Gewerbestrasse 11, 6330 Cham, Switzerland

Melissa Lukings
For my family, friends, and all of those who have supported me throughout my journey.
And for my cat children,
my shadow goblins,
my floofy beneficiaries,
my little meow nuggets... with limitless love.

Arash Habibi Lashkari
For my wife Farnaz and children, Kourosh and Kianna,
And my father Bahman, mother Zeynab and sister Ziba,
And my teachers and lecturers, for all the lessons you've taught me.

Preface

This book is one piece of the Understanding Cybersecurity Series (UCS) research programme, which will produce a varied collection of cybersecurity resources for researchers and readers of all backgrounds. Our first online article series in this piece, entitled "Understanding Canadian Cybersecurity Laws", is a 10-part series which was published online between January 2020 and April 2021.

- Understanding Canadian Cybersecurity Laws: The Foundations
- Understanding Canadian Cybersecurity Laws: Privacy and Access to Information, the Acts
- Understanding Canadian Cybersecurity Laws: Privacy Protection in the Modern Marketplace—PIPEDA
- Understanding Canadian Cybersecurity Laws: Interpersonal Privacy and Cybercrime—Criminal Code of Canada
- Understanding Canadian Cybersecurity Laws: "Insert Something Clever Here"—Canada's Anti-Spam Legislation
- Understanding Canadian Cybersecurity Laws: Peer-to-Peer Privacy Protection—"Intrusion Upon Seclusion" and the Protection of Intimate Images
- Understanding Canadian Cybersecurity Laws: Deep, Dark and Undetectable—Canadian Jurisdictional Considerations in Global Encrypted Networks
- Understanding Canadian Cybersecurity Laws: Measuring Up—Outlining Existing Federal Cybersecurity Legislation in Canada, the UK, Australia and the United States
- Understanding Canadian Cybersecurity Laws: Legislative Modernisation—Responding and Adapting to Technological Change in a Global Domain
- Understanding Canadian Cybersecurity Laws: Refactored—Our Series in Summary

Flowing from the success of this first article series, we were recognised with a Gold Medal for Best Blog Column in the Business Division at the 2020 Canadian Online Publishing Awards, which was remotely held in 2021 as a result of the ongoing COVID-19 pandemic. The research continued with the publication of our first book, "Understanding Cybersecurity Law and Digital Privacy: A Common Law Perspective", which was published by Springer Nature Switzerland AG on 9 December 2021, with hard copies of the book being distributed beginning in January 2022.

Starting in 2022, we began our second online series, "Understanding Current Cybersecurity Challenges in Law", which contains six parts and addresses many of the emerging trends and larger legal issues pertaining to cybersecurity around the world, including determination of digital jurisdictional authority, user-generated digital content ownership and other topics.

This book represents the results of our work to bring a nuanced and comprehensive understanding of current cybersecurity challenges and the law to the greater community and increase public awareness of these important issues in our rapidly changing digital world.

Toronto, ON, Canada

Melissa Lukings
Arash Habibi Lashkari

Introduction

Digital governance refers to the strategic management of a governing body or corporation, which is connected to their position and function online—including the impacts of their actions in that position. **Digital corporate social responsibility** extends the social obligations related to the digital governance of a corporation or other governing body to that which is connected to their position within the digital sphere. The legal issues which intersect with, and relate to, user-generated content can be wide-ranging, including determination of ownership and copyright, personal privacy protection, individual freedom of expression, issues of cyberbullying, illegal content distribution and large-scale corporate data breaches.

Cloud computing and cloud-service providers have brought about a significant shift in the globalised delivery of information technology services and computing resources. When data is stored in a cloud environment, regardless of where the cloud-service resources are physically located, the data may be subject to the laws of other countries. In law, data sovereignty refers to the control a nation—or state—has over access to and disclosure of its digital information, and subject only to its laws. Where a cloud-service provider—as a large-scale corporate entity—is operating on a global level, it could end up being required to comply with a court order, warrant or subpoena request from a foreign law enforcement agency seeking to obtain data relating to the affairs of another nation.

When a crime takes place or a dispute occurs between two parties who reside within the same geographic location, the determination of which law applies to that scenario is easily distinguished as being the law of that geographic location. When a dispute occurs between two parties who reside in different nations—or legal jurisdictions—the methodology for determining which country's laws and legal forum should apply is rooted in the topics of private international law, international criminal law and conflicts of law. While these topics are by no means straightforward or simple, the strategy for addressing legal disputes or criminal offences which occur between or across legal jurisdictions becomes greatly complexified when the alleged crime, dispute or other offence takes place online or is otherwise enabled via the Internet.

While jurisdiction is often linked to sovereignty over a territorial location, jurisdiction can also exist without a connection to territory. The type of jurisdictional authority held by a governing body indicates whether that nation or state can undertake enforcement action to uphold its law. This becomes more complex when we consider online activities, particularly cybercrimes, defamation and other tortious activities, as there may not necessarily be a defined geographic area to distinguish which governing authority has the legal jurisdiction to address these matters.

Acknowledgement

For Mahsa Amini and all of those fighting for Women, Life, and Freedom.

Contents

About the Authors

Melissa Lukings received her Juris Doctor (JD) from the Faculty of Law at the University of New Brunswick (UNB) and her Bachelor of Arts (BA) in Linguistics from the Memorial University of Newfoundland (MUN). She is a published author, content writer, intersectional activist and research assistant with a background in social justice, human rights, advocacy, data privacy and cybersecurity law.

Melissa has an extensive history of social activism and community involvement with non-profit organisations, particularly those with a primary mandate that centres on: harm reduction; mental health advocacy, education and awareness; crisis and suicide intervention; violence prevention; sexual health and reproductive justice; human rights and access to justice within gender-based marginalised community networks.

While completing her law degree at UNB—specifically between 2019 and 2021—Melissa Lukings co-authored a ten-article series on cybersecurity laws in Canada, the "Understanding Canadian Cybersecurity Laws" series, which was published with IT World Canada. This series was honoured with the receipt of the Gold Award for Best Blog Column in the Business Division of the 2020 Canadian Online Publishing Awards, which was held remotely in early 2021 to accommodate for the ongoing and fluctuating restrictions of the COVID-19 pandemic. Following the success of the article series, she co-authored a textbook, "Understanding Cybersecurity Law and Digital Privacy: A Common Law Perspective" which was published by Springer Nature AG Switzerland and released in 2022.

During this same period of time, Melissa was invited to appear as an expert witness to testify in the Parliament of Canada; twice before the House of Commons' Standing Committee on Access to Information, Privacy and Ethics (ETHI) and once before the House of Commons' Standing Committee on Justice and Human Rights (JUST). These appearances were made during the preliminary study and consultation regarding current online privacy challenges as well as during the parliamentary review of the Criminal Code provisions which were enacted by the 2014 implementation of the PCEPA.

In mid-2022, between the completion of the requirements for her law degree and the commencement of her articling term, Melissa co-authored a second article series, called "Understanding Current Cybersecurity Challenges in Law" and consisting of six parts, also published by IT World Canada.

At the time of this publication, Melissa Lukings continues to work on a handful of research projects, covering a wide variety of topics which range from cybersecurity and data privacy law to legal reform and access to justice within the Canadian judicial system. She currently lives in Ottawa, ON, where she is working on completing her articling term, as required for legal licensing with the Law Society of Ontario (LSO).

Arash Habibi Lashkari is a Canada Research Chair (CRC) in Cybersecurity. He is a senior member of the IEEE and an Associate Professor at York University (Canada). Prior to this, he was an Associate Professor at the Faculty of Computer Science, University of New Brunswick (Canada), and the Research Coordinator of the Canadian Institute for Cybersecurity (CIC). His research focuses on cyber threat modeling and detection, malware analysis, big data security, internet traffic analysis, and cybersecurity dataset generation.

Arash Lashkari has over 22 years of teaching experience, spanning several international universities, and was responsible for designing the first cybersecurity Capture the Flag (CTF) competition for post-secondary students in Canada. He has been the recipient of 15 awards at international computer security competitions - including three gold awards - and was recognized as one of Canada's Top 150 Researchers for 2017. In 2020, Dr. Lashkari was recognized with the University of New Brunswick's prestigious Teaching Innovation Award for his personally created teaching methodology, the Think-Que-Cussion Method.

He is the author of ten published books and more than 100 academic articles on a variety of cybersecurity-related topics and the co-author of the national award-winning article series, "Understanding Canadian Cybersecurity Laws," which was recently recognized with a Gold Medal at the 2020 Canadian Online Publishing Awards.

List of Figures

List of Tables

Chapter 1
Data Sovereignty

At first glance, the legal concept of sovereignty can seem unnecessarily complex. This complexity becomes magnified when we apply the concept of sovereignty to the intangible concept of data.

Data sovereignty is the idea that the laws and governance structures of a nation extend in application to the data collected within that nation. It is closely linked to the topics of data security, cloud computing and storage and technological sovereignty. Unlike other forms of sovereignty, data sovereignty is specifically concerned with questions surrounding the data in and of itself.

As an example, with respect to Canada, **data sovereignty** refers to Canada's right to control the access to, and disclosure of, its digital information and subject only to Canadian laws. The Government of Canada, in their White Paper on *Data Sovereignty and Public Cloud*, outlines the governmental strategy whereby cloud services are identified and evaluated as the principal delivery option when initiating information technology investments, initiatives, strategies and projects. The White Paper outlines the "cloud-first" strategy selected by the Government of Canada and recommends that departments and agencies consider various cloud deployment models in the following order of priority:

1. Public cloud
2. Hybrid cloud
3. Private cloud
4. Non-cloud storage options[1]

[1] https://www.canada.ca/en/government/system/digital-government/digital-government-innovations/cloud-services/gc-white-paper-data-sovereignty-public-cloud.html

M. Lukings, A. Habibi Lashkari, *Understanding Cybersecurity Law in Data Sovereignty and Digital Governance*, Progress in IS,
https://doi.org/10.1007/978-3-031-14264-2_1

1.1 Topic Overview

In this chapter, we will build the foundations of contextual knowledge which will be necessary to tackle the chapters to follow. To begin, we will define the concept of data in both the information technology context and the legal context. We will outline the primary methods of data storage and introduce the intersectional concepts of data residency, data localisation and data sovereignty in law. Finally, we will wrap up this chapter by exploring the legal considerations which apply to sovereignty—specifically data sovereignty—in practice.

1.2 Defining Data in Context

The term **data** encompasses many things. Data can be textual, numerical, audial or visual. It can be qualitative or quantitative, public, private or personal. Data can be a collection of facts, composed of numbers, words, measurements, observations or descriptions of things. Data can also refer to distinct pieces of information, usually formatted and stored in a way that is concordant with a specific purpose.

In computing, data is information that has been translated into a form that is efficient for movement or processing. Relative to today's computers and transmission media, data is information converted into binary digital form. It is acceptable for data to be used as a singular subject or a plural subject. The term "raw data" is used to describe data in its most basic digital format.

The concept of data in the context of computing has its roots in the work of Claude Shannon, an American mathematician known as the father of information theory. He ushered in binary digital concepts based on applying two-value Boolean logic to electronic circuits. Binary digital formats underlie the CPUs, the semiconductor memories and the disk drives, as well as many of the peripheral devices which have become common modern computing.

Computers represent data, including video, images, sounds and text, as binary values. Early computer input for both control and data took the form of punch cards, followed by magnetic tape and the hard disk. Quantities of stored data can be measured in bits and bytes. A "bit" is the smallest unit of data and represents just a single value, whereas a "byte" consists of eight binary digits. Data storage and memory systems can be measured in megabytes and gigabytes. Indeed, as the amount of data which is collected and stored continues to grow, the units of data measurement must also grow to accommodate this level of expansion.

Data also has meaning beyond its use in computing applications oriented towards data processing. For example, in electronic component interconnection and network communication, the term data is often distinguished from "control information", "control bits" and similar terms to identify the main content of a transmission unit. Moreover, in science, the term data is used to describe a gathered body of facts, as is the case in fields such as finance, marketing, demographics, research and healthcare.

Fig. 1.1 Data word cloud

Early on, the importance of data in business computing became apparent as evidenced by the popularity of the terms "data processing" and "electronic data processing", which eventually came to encompass the field of information technology. Over the history of corporate computing, specialisation occurred, and a distinct data profession emerged along with the growth of corporate data processing. Greater specialisation was developed as databases, database management systems and then relational database technology became a necessity to organise the data collected and stored. The growth of the Internet and smartphones over the past decade has also led to a surge in digital data creation. Data, as we now know it, includes text, audio and video information, as well as log and web activity records, and other content—much of it unstructured.

The term "big data" has been used to describe data in the petabyte range or larger. As web-based e-commerce has spread around the world, big data-driven business models have evolved which treat data as an asset. Such trends have also spawned greater preoccupation with the social uses of data and data privacy among the general population.

1.2.1 Types and Features

The different types and features of data are illustrated as a "word cloud" in Fig. 1.1.

As the cloud graphic shows us, data itself can have many different and overlapping qualities: from qualitative data to quantitative data and from text to numbers, to binary representations of data to boolean data. It can involve many different forms of content, information and material with an equally wide range of descriptive features and qualities. As with physical, non-digital, forms of information, data can have ownership implications. When data is considered to be owned by an entity or individual, that data is considered to be "property".

1.2.2 *Data as Property*

There is an ongoing discussion within the legal and technological communities regarding whether or not data should be considered to be "property" in law.

Property, as a concept in law and political economics, is any physical or intangible entity that is owned by a person, jointly by a group of people, or a legal entity such as a corporation or society.[2] Property can also be one or more components or a greater thing, such as the pieces that make up a person's estate. Depending on the nature of the property, the owner of the property has the right to consume, sell, rent, mortgage, transfer, exchange or destroy their property, as well as the right to exclude others from doing these things. Property can be either moveable or immovable property, tangible or intangible property.[3]

Immovable property, or **real property**, refers to the physical property consisting of land and all structures—also called fixtures or improvements—integrated with or affixed to that land, including crops, buildings, machinery, wells, dams, ponds, natural resources, mines, canals and roads, among other things.[4] The term "real property" arises from the now-discontinued "form of action" system for handling legal claims, which distinguished between real property disputes and personal property disputes.[5]

Movable, or **personal**, property was, and continues to be, all property that is not real property. In common law legal systems, **personal property** is also called chattels or personality.[6] In civil law systems, personal property is often called movable property or movables—that is, any property that can be moved from one

[2]Baron, J. B. (2013). Rescuing the bundle-of-rights metaphor in property law. U. Cin. L. Rev., 82, 57.

[3]von Benda-Beckmann, F., von Benda-Beckmann, K. and Wiber, M.G., 2006. The properties of property. Changing properties of property, 40, pp. 1–39.

[4]Niles, R.D., 1933. The Rationale of the Law of Fixtures: English Cases. NYULQ Rev. 11, p. 560.

[5]Hetland, J.R., 1965. Real Property and Real Property Security: The Well-Being of the Law. Cal L. Rev., 53, p. 151.

[6]O'Keefe, K.M., 1983. The Classification Issue and the Law of Fixtures: A Chattel by Any Other Name... J. St. Tax'n, 2, p. 37.

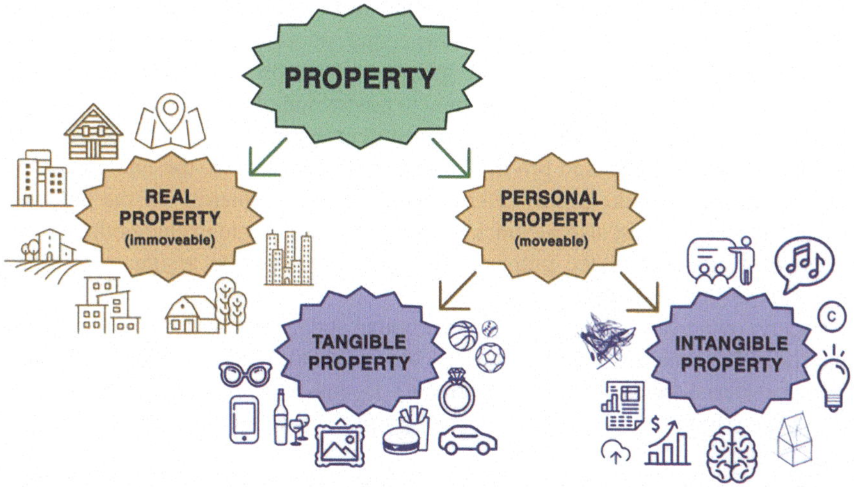

Fig. 1.2 Property types

location to another.[7] Personal property can be contrasted with real property or immovable property, that being, land and the fixtures attached to the land.[8] Personal property can be further subdivided into two categories: tangible personal property and intangible personal property.[9]

Tangible personal property includes physical things that can be moved, touched or felt and includes physical objects like household items, furniture, books and other moveable physical things.[10] Tangible property is often what people think of when we discuss personal possessions.[11]

In contrast to tangible property, **intangible personal property** is moveable property that lacks a physical substance—that is, it cannot be physically picked up and moved, touched or felt—but which still represents something of value.[12] This can include patents, copyrighted material, franchises, registered trademarks, trade names, software, research data and other non-physical assets.[13] This is illustrated in Fig. 1.2.

[7] Rahmatian, A., 2008. A comparison of German moveable property law and English personal property law. J. Comp. L., 3, p. 197.

[8] Niles, R.D., 1934. The Intention Test in the Law of Fixtures. NYULQ Rev., 12, p. 66.

[9] Moore, A.D., 1998. Intangible property: privacy, power, and information control. American Philosophical Quarterly, 35(4), pp. 365–378.

[10] Carnahan, W., 1934. Tangible Property and the Conflict of Laws. U. Chi. L. Rev., 2, p. 345.

[11] Arezzo, E., 2007. Struggling Around the Natural Divide: The Protection of Tangible and Intangible Indigenous Property. Cardozo Arts & Ent. LJ, 25, p. 367.

[12] Child, J.W., 1990. The moral foundations of intangible property. The Monist, 73(4), pp. 578–600.

[13] Hardy, I.T., 2000. Not So Different: Tangible, Intangible, Digital, and Analog Works and Their Comparison for Copyright Purposes. U. Dayton L. Rev., 26, p. 211.

Property law refers to the application of the laws which govern the relationships between individuals and possessions.[14] Ownership, in property law, refers to the set of legal rights—held by an individual, group, corporation or government—which grant the legal owner of a property: the right to possession of a property, the privilege of use of that property and the right of control over that property.

These collective rights and privileges—the ability to possess, use and control or transfer the property—are jointly referred to as the "legal interest" of the property owner. The legal interest also includes the right of the owner to exclude others from the use or enjoyment of the property, as well as the right to assign or reassign possession of that property.

Property ownership can apply to both corporeal and incorporeal things. Corporeal property is that which is tangible, such as objects, land, vehicles, books and other physical material items. Incorporeal property is that which is intangible, such as patents, copyrights, trademarks, intellectual property and other non-physical immaterial things.

But where does our modern concept of electronic data fit into all of this? Can data be property that is owned by an individual or entity? Whose data, is it?

As expected, with the rise of cloud computing, many countries have passed various laws around control and storage of data, all reflecting various measures of data sovereignty.[15] Accordingly, many countries now have some sort of data sovereignty laws in place to govern the ownership and control of data collected within that country.[16]

1.2.3 Ownership and Interest

It is easy to understand the appeal of treating our data as our property. It touches on what the foundational privacy thinker, Alan Westin, identified as an essential aspect of privacy, a right "to control, edit, manage, and delete information about [individuals] and decide when, how, and to what extent information is communicated to others".[17] It expresses the unfairness people feel about an inherently asymmetrical marketplace in which we know little about the data we share but the companies that receive the data can profit by extracting marketable information.

[14] Alexander, G. and Donahue, C. (2018 January 25). Property law. Encyclopaedia Britannica. https://www.britannica.com/topic/property-law

[15] Irion, Kristina (2012-12-01). "Government Cloud Computing and National Data Sovereignty". *Policy & Internet*. **4** (3–4): 40–71.

[16] "Gilmore, David, DataFleets, 'Google Scrapped Cloud Initiative in China, Other Markets', *Bloomberg News*". 8 July 2020.

[17] Westin, A. F. (1967). *Privacy and freedom* [first ed.] New York: Atheneum.

Indeed, as recently as 2018, Mark Zuckerberg—while appearing in widely observed US congressional hearings—iterated, more than a few times, that "people own all of their own content" on Facebook.[18] Herein lies the issue.

You see, it is not your data, but it is not their data either. If we opt to treat data like it is property, we may fail to recognise the value that different varieties of personal information can serve or the abiding interest that individuals have in their own personal information. This applies even if an individual chooses to "sell" their data. Data is not a commodity. It is information. Any system of information rights—whether patents, copyrights and other intellectual property, or privacy rights—presents some tension when coming up against the strong socio-political interest in the free flow of information that we can see reflected in many constitutional and other statutory provisions. It is precisely because this data has value to others, and to society across a myriad of uses, that our individual personal information is in demand.

Treating personal information as property to be licensed or sold may also serve to induce people to trade away their privacy rights for very little value while still injecting enormous amounts of friction into the free flow of information. The better way to strengthen privacy is to use a strategy which would ensure that individual privacy interests are respected as personal information flows to desirable uses, rather than a strategy which reduces personal data to the status of a mere commodity.

To unpack the private and public interests inherent in personal information, we can consider the simple case of a person's name. It is a fundamental part of one's individual identity, and it is the first item on any list of "personally identifiable information" protected by data privacy rules. We each have vital interest in our names and the assets and identities that we link to those names.

Yet this interest is not exclusive. Our names are also a way that others recognize us, associated by friends and family with who and what we are, and used by society for voting rolls, property registries, financial accounts and countless other social, economic and civic contexts. This makes our names essential social and economic currency that is valuable to others, and it is because of this social and economic significance that naming is regulated. We are assigned names at birth, registered on birth certificates. It takes official approval to change these names. In this way, as personal as our identities are to us, they are also instruments of a well-functioning society.

Other personal information that we necessarily share also involves intersecting interests. The transactions we conduct through a bank become part of the business records of the bank, but the bank's interest in these records does not necessarily mean it has absolute control to do what it wants with this information.[19] Indeed, courts in the United States have found that records of cell phone calls and other business records can be obtained from the parties that hold these records.

---

[18] https://www.washingtonpost.com/news/fact-checker/wp/2018/04/13/fact-checking-mark-zuckerbergs-testimony-on-facebook-and-data-collection/

[19] *United States v Miller* (1976) 425 US 435.

The same intersecting interests apply to much of the other information we share as we interact with digital services, such as e-commerce transactions, mobile applications, messaging services and other communications, and pictures or information that we share on our social networks, among others.

Of course, much of the data that we generate is shared more widely than just with recipients like these, but here, too, the interests in information from the data are not exclusive. The digital economy operates in broad ecosystems of data sharing for diverse uses. Some of the more well-accepted examples are for credit reporting, bank clearing and compliance with know-your-customer rules, fraud monitoring and security protocols.

Many of these functions are recognised explicitly in existing laws and standards, but these can vary in both jurisdiction and in scope, from the first US federal-level privacy law in 1970—the *Fair Credit Reporting Act (FCRA)*[20]—to the much more recent implementation of the regional *General Data Protection Regulation* (GDPR)[21] implemented by the European Union and the state-based *California Consumer Privacy Act.*[22] There are very few sectors that exist entirely outside of such legal systems. As businesses continue to outsource functions and move data processing onto cloud-based services, and as remote sensing of all kinds begins to deploy across environments, an ever-increasing amount of our personal data and information will exist in the possession of entities which may be entirely unknown to us, as the consumers.

1.2.3.1 Example: United States

We can see an example of this in a 2017 US case, which was decided in 2018. In this case, the court required a warrant for cell phone location data that service providers retain for their use while also recognising the expectation of individual phone users to retain privacy with regard to any location-based data generated from their phones.

---

[20] *The Fair Credit Reporting Act*, 15 U.S.C. § 1681 et seq, was enacted as US federal government legislation to promote the accuracy, fairness and privacy of consumer information contained in the files of consumer reporting agencies.

[21] *EU General Data Protection Regulation (GDPR):* Regulation (EU) 2016/679 of the European Parliament and of the Council of 27 April 2016 on the protection of natural persons with regard to the processing of personal data and on the free movement of such data and repealing Directive 95/46/EC (General Data Protection Regulation), OJ 2016 L 119/1.

[22] *California Consumer Privacy Act.* Section 3, Title 1.81. 5 of the CCPA, added to Part 4 of Division 3 of the California Civil Code.

1.2.4 Data Privacy

Owners of data repositories collect information from various data suppliers and store it for various purposes. Unfortunately, once the supplier releases their data to the collector, they must rely upon the collector to use it for the purposes for which it was provided. Over the last 20 years, the advancements in information technology (IT) have radically changed our lives. We can access a variety of services and information anywhere anytime using our personal computers, mobile phones, tablets or any device with an Internet connection. Although this situation has clearly brought enormous benefits to our society, the development of information technology has also had a significant impact on users' privacy. As a matter of fact, more and more personal information is collected, processed, shared and disseminated. This includes demographic data, medical data, tweets, emails, photos, videos as well as location information. There are a variety of reasons for collecting, sharing and disseminating personal information. For instance, public, private and governmental organisations might disclose or share their data collections for research or statistical purposes, for providing services more efficiently and effectively or because forced by laws and regulations. However, disseminating and sharing personal information may put individuals' privacy at risk: How should personal information be collected and processed? How should privacy be defined and enforced?

Data privacy or information privacy is a part of the data protection area that deals with the proper handling of data focusing on compliance with data protection regulations.

Data privacy is centred around how data should be collected, stored, managed and shared with any third parties, as well as compliance with the applicable privacy laws such as the *California Consumer Privacy Act (CCPA)*[23] or the *General Data Protection Regulation*[24] *(GDPR)*.[25,26]

The privacy definitions (and corresponding data protection techniques) can be broadly classified in the following two main categories[27]:

[23] *California Consumer Privacy Act.* Section 3, Title 1.81. 5 of the CCPA, added to Part 4 of Division 3 of the California Civil Code.

[24] *EU General Data Protection Regulation (GDPR):* Regulation (EU) 2016/679 of the European Parliament and of the Council of 27 April 2016 on the protection of natural persons with regard to the processing of personal data and on the free movement of such data and repealing Directive 95/46/EC (General Data Protection Regulation), OJ 2016 L 119/1.

[25] "CCPA/California Consumer Privacy Act", https://dataprivacymanager.net/glossary/ccpa-california-consumer-privacy-act/

[26] "What is General Data Protection Regulation (GDPR)?", https://dataprivacymanager.net/what-is-gdpr-general-data-protection-regulation/

[27] Sabrina De Capitani di Vimercati and Sara Foresti and Giovanni Livraga and Pierangela Samarati, "Data Privacy: Definitions and Techniques", Int. J. Uncertain. Fuzziness Knowl. Based Syst., Vol. 20, pp. 793–818, 2012.

- **Syntactic privacy definitions** capture the protection degree enjoyed by data respondents with a numerical value. Data protection techniques falling in this category are aimed at satisfying a syntactic privacy requirement (e.g. each release of data must be indistinguishably related to no less than a certain number of individuals in the population).
- **Semantic privacy definitions** are based on the satisfaction of a semantic privacy requirement. Data protection techniques falling in this category are aimed at satisfying a property that must be satisfied by the mechanism chosen for releasing the data (e.g. the result of an analysis carried out on a released dataset must be insensitive to the insertion or deletion of a tuple in the dataset).

We can illustrate the basic concepts of these two definitions and related data protection techniques.

1.2.4.1 Syntactic Data Protection Techniques

Data to be protected are typically released in the form of a table (microdata table) defined on a set of attributes that can be classified as follows:

- Identifiers: attributes that uniquely identify a respondent (e.g. SSN)
- Quasi-identifiers (QI): attributes that, in combination, can be linked with external information to re-identify (all or some of) the respondents to whom information refers or reduce the uncertainty over their identities (e.g. DoB, sex and zip)
- Confidential attributes: attributes that represent sensitive information (e.g. disease)
- Non-confidential attributes: attributes that are not considered sensitive by the respondents and whose release is harmless (e.g. favourite colour)

Syntactic data protection techniques are based on the assumption that the release of a microdata table can put at risk only the privacy of those individuals contributing to the data collection.[28] The first step for protecting their privacy consists of removing (or encrypting) explicit identifiers before releasing the table. However, a de-identified microdata table does not provide any guarantee of anonymity, since the quasi-identifier can still be linked to publicly available information to re-identify respondents.

[28] The word "syntax" comes from the Ancient Greek for "coordination" or "ordering together". In spoken and written language, syntax refers to the set of rules that determines the arrangement of words in a sentence. In linguistic communication, syntactic rules help readers and writers make sense of sentences. Along with diction, it is one of the key ways writers convey meaning in a text, by creating various rhetorical or literary effects.

1.2.4.2 Semantic Data Protection Techniques

Semantic data protection techniques have recently been proposed to protect the privacy of both data respondents and individuals who are not included in data undergoing public release.

Semantic techniques operate in the following two scenarios[29]:

1. Non-interactive scenario:
   - This consists of the release of a data collection.
   - Protection techniques are therefore used to compute a privacy-preserving dataset, which is representative of the original data collection.
2. Interactive scenario.
   - This consists of evaluating queries over a private data collection managed by the data holder, without revealing to the requesting recipient any information that is not intended for disclosure.
   - Protection techniques are used to guarantee that the query result (also when possibly combined with other results collected by data recipients) cannot be exploited to gain information that should be kept secret.

While syntactic techniques traditionally guarantee data protection preserving the truthfulness of the released information, semantic techniques typically add noise to the released data. Noise addition perturbs the original content of the dataset, thus achieving privacy at the price of truthfulness.

1.2.5 Data Security

Data security refers to the process of protecting data from unauthorised access and data corruption throughout its life cycle. This includes protecting your data from attacks that can encrypt or destroy data, such as ransomware, as well as attacks that can modify or corrupt your data. Data security also ensures data is available to anyone in the organisation who has access to it.

There are several data security solutions such as:

- **Cloud Data Security**—Protection platform that allows you to move to the cloud securely while protecting data in cloud applications.
- **Data Encryption**—Data-centric and tokenisation security solutions that protect data across enterprise, cloud, mobile and big data environments.

---

[29] Semantics is the study and analysis of how language is used figuratively and literally to produce meaning. Semantics seeks to describe how words are used not to prescribe how they should be used. The word is derived from "sema", a Greek word meaning signs. Semantic barriers to communication are the symbolic obstacles that distort the message in some other way than intended, making the message difficult to understand.

Fig. 1.3 In-depth, onion model for defence

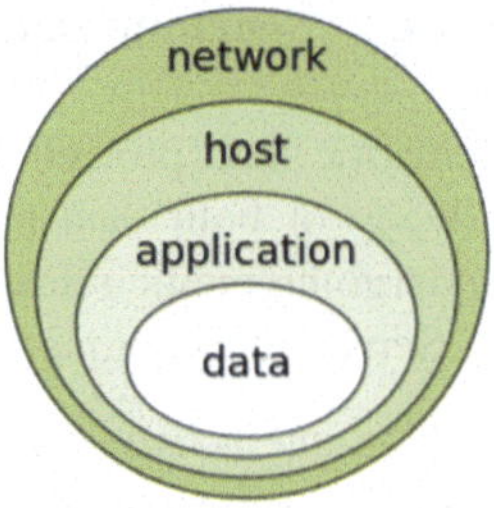

- **Hardware Security Module**—Hardware security module that guards financial data and meets industry security and compliance requirements.
- **Key Management**—Solution that protects data and enables industry regulation compliance.
- **Enterprise Data Protection**—Solution that provides an end-to-end data-centric approach to enterprise data protection.
- **Payments Security**—Solution provides complete point-to-point encryption and tokenisation for retail payment transactions, enabling Payment Card Industry (PCI) scope reduction.
- **Big Data, Hadoop and IoT Data Protection**—Solution that protects sensitive data in the data lake, including Hadoop, Teradata and other big data platforms.
- **Mobile App Security**—Protecting sensitive data in native mobile apps while safeguarding the data end to end.
- **Web Browser Security**—Protects sensitive data captured at the browser, from the point the customer enters cardholder or personal data and keeps it protected through the ecosystem to the trusted host destination.
- **eMail Security**—Solution that provides end-to-end encryption for email and mobile messaging, keeping personally identifiable information and personal health information secure and private.

Figure 1.3 shows a generic in-depth view of onion model of defence where data is layered inside along with the application adopted, its host and network. It shows a powerful encrypted network, reliable host and application control can ensure the security and safety of your data from unauthorised access and misuse.[30]

Donn B. Parker defines the six main elements of data security known as Parkerian hexad. And the elements of Parkerian hexad are[31]:

- **Confidentiality**—Confidentiality refers to limits on who can get what kind of information. For example, executives concerned about protecting their enterprise's strategic plans from competitors; individuals are concerned about unauthorised access to their financial records.

[30] Sandeep Dhawan, "Information and Data Security Concepts, Integrations, Limitations and Future", International Journal of Advanced Information Science and Technology (IJAIST), Vol.1, No.2, October 2014.

[31] Donn B. Parker, "Computer Security Handbook", Fifth Edition, Wiley, 2009.

- **Control/Possession**—Utility means usefulness. For example, suppose someone encrypted data on disk to prevent unauthorised access or undetected modifications—and then lost the decryption key—that would be a breach of utility.
- **Integrity**—Integrity means that the information is whole, sound and unimpaired (not necessarily correct). It means nothing is missing from the information; it is complete and in intended good order.
- **Authenticity**—Authenticity refers to the veracity of the claim of origin or authorship of the information. For example, one method for verifying the authorship of a handwritten document is to compare the handwriting characteristics of the document to a sampling of others which have already been verified.
- **Availability**—Availability means having timely access to information. For example, a disk crash and denial-of-service attacks both cause a breach of availability. Any delay that exceeds the expected service levels for a system can be described as a breach of availability.
- **Utility**—Suppose a thief were to steal a sealed envelope containing a bank debit card and its personal identification number. Even if the thief did not open that envelope, it's reasonable for the victim to be concerned that the thief could do so at any time.[32]

1.2.6 Corporate Data/Media Content

Digital content refers to any type of content which can be transmitted online or via networked computer systems. Digital content—also known as digital media—comes in many forms, from text, audio, and video files to graphics, animations, programs and images.[33] Digital content can be online—via the Internet—or offline, such as content stored on a USB flash drive. Online digital content comes from a wide range of sources, from large media companies to small businesses to content entrepreneurs to social media account users.

Most generally speaking, digital content refers to information which is made available for download or distribution on electronic media, such as an ebook, mobile game, video or audio file. However, many creative content and online media producers have argued that the term "digital content" should extend to encompass anything that can be published.[34] Table 1.1 below provides an outline of some of the major varieties of digital content and their descriptions.

It follows, then, that every tweet, every hashtag, every status update, every video upload, every blog update and every social media share, retweet, reblog, upload, etc. will all fall under the umbrella of "digital content" as it applies in this context. When

[32] Donn B. Parker, "Computer Security Handbook", Fifth Edition, Wiley, 2009.

[33] Consider: Social media posts are online digital content.

[34] The authors support the position that digital content includes anything that can be published, rather than limited to that which is downloaded or distributed on electronic media.

Table 1.1 Digital content types and descriptions

Type of content	Description of digital content
Ad/advertisement	Online advertising, also known as Internet advertising, digital advertising or web advertising, is a form of marketing and advertising which uses the Internet as a medium to contain website traffic and deliver promotional marketing messages to targeted consumers. Online ads can take the form of display and video ads, social media ads, commercial electronic messages, push notifications, mobile advertising and other forms.
App/application	An app, which is short for "application", is a type of software that can be installed and run on a computer, tablet, smartphone or other electronic devices. An app most frequently refers to a mobile application or a piece of software that is installed and used on a computer.
Article	A piece of writing, usually nonfiction and on a specific topic, forming an independent part of a book, newspaper, magazine, journal or other publication.
Blog/weblog	A blog, which is short for "weblog", is a discussion, reflection or informational website published on the Internet and consisting of discrete, often informal journal or diary-style textual entries, called posts.
Comic/cartoon	These are mediums used to express ideas with images, often combined with text or other visual information contained in speech bubbles or captions or conveyed through other textual devices. Comics typically take the form of a sequence of panels of images whose size and arrangement contribute to narrative pacing.
Data archive	Data archiving is the process of collecting older data that is no longer actively used and moving it to a separate secure storage location or device for long-term retention. A data archive is the storage place for data that is important but does not need to be accessed or modified regularly—or at all—so that it can be retrieved if needed.
Digital audio file	A method of preserving sound in digital form—that is, the form in which audio signals are digitised and transformed into a series of pulses that correspond to patterns of binary digits and are recorded as such on the surface of a magnetic tape or optical disc.
E-book	An e-book, also known as an ebook or eBook, is a downloadable book publication made available in digital form, consisting of text, images or both, which is readable on the flat-panel display of computers or other electronic handheld devices. They can consist of just the electronic text or may contain extra features, such as audio, video or hyperlinks.
GIF/graphical interchange format	A GIF, which stands for graphical interchange format, is a bitmap image format consisting of a series of images or soundless video that will loop continuously without the user clicking anything.
Infographic	An infographic, a portmanteau derived from "information" and "graphic", that is, a means of conveying information through a visually appealing collection of imagery, charts and minimal text. Infographics provide a quick way to synthesise data, distilled in a way that is accessible and easy to digest and presented in a clear and uncomplicated manner.

(continued)

Table 1.1 (continued)

Type of content	Description of digital content
Livestream	Livestreaming refers to online streaming media simultaneously recorded and broadcast in real time—live and uncensored. It is often referred to simply as streaming, but this abbreviated term is ambiguous because "streaming" may refer to any media delivered and played back simultaneously without requiring a completely downloaded file.
Meme	A meme is a virally transmitted image embellished with text, usually sharing pointed commentary on cultural symbols, social ideas, systems of behaviour or current events. A meme is typically a photo or video, although sometimes it can be a block of text.
Podcast	A podcast is an episodic series of spoken word digital audio files. These files are made available for download by followers and subscribers of the podcast's channel.
Social media post	Content shared on social media through a user's profile. It can be as simple as a blurb of text, but can also include images, videos and links to other content. Other users of the social network can like, comment and share the post.
User-generated content (UGC)	User-generated digital content can consist of any form of content: written, videoed, posted, blogged, created and uploaded to a forum by individual people rather than commercial brands, organisations or companies. It can include images, videos, text and/or audio that has been posted by the user on online platforms.
Webinar	A webinar is an online seminar that turns a presentation into a real-time conversation from anywhere in the world. This allows a speaker from the hosting organisation or company to share PowerPoint presentations, videos, web pages or other multimedia content with audiences that can be located anywhere.

referring to digital content which is interpreted through a commercial marketing perspective, the term "media" may also be used interchangeably with "content".

Corporations which operate online, or have any form of online presence, can stand to benefit from a variety of types of digital content—known in these circles as digital media. Online digital media can be further broken down into three categories: (1) earned media, (2) owned media and (3) paid media. The relationship between the corporation and the media content is given in the name of the type of media.

Earned media refers to content publicity gained through promotional efforts other than paid media advertising.[35] Earned media does not include any media content which is bought, owned, managed or directly generated by the corporation, brand or organisation. Rather, it must be grown organically, as content related to that entity receives gradual recognition, consumer loyalty and/or an online following

[35] Yu, Jim (7 October 2013). "Earned Media Rising - The Earned Media Ripple Effect". *Column: Social Media Marketing Column*. Marketing Land.

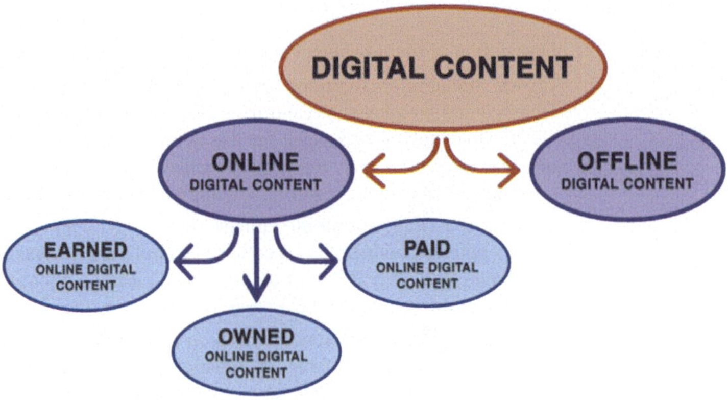

Fig. 1.4 Subsets of online digital content

through online communication channels. Common examples of earned media include social media shares, positive consumer reviews and online referrals.[36]

Owned media is a type of advertising or media presence which is generated, managed and/or otherwise controlled by the brand or company. Owned media also includes communication channels that are within the control of an entity, such as its own websites, blogs or email accounts.

Paid media, or paid advertising, refers to external marketing efforts that involve a paid placement, such as pay-per-click advertising, branded content and display ads. It is often thought of in the context of traditional advertising strategies, which gained substantial popularity prior to the acceleration and widespread adoption of the Internet and online commercial activity.

The graphic provided below (Fig. 1.4) shows the division of online digital content into these three distinct subtypes. While offline digital content is certainly important, as we are not considering offline digital content in this discussion, the subsets of that branch have not been included in the graphic.

These three types of media can exist both on and offline; however, for the purposes of this writing, we will consider digital media content in the context of online communication channels.

It should be noted that there are three distinguishing features which separate these three types of media: connection, creation and control.

1. **Connection** refers to the relationships between the content creator and the company or other corporate entity.
2. **Creation** is indicative of the source of the content.
3. **Control** refers to the level of management in terms of the amount of power over the execution and follow-up of the content which is held by the company.

[36] Stephen, A.T.; Galak, J. (October 2012). "The Effects of Traditional and Social Earned Media on Sales: A Study of a Microlending Marketplace". *Journal of Marketing Research*. 49 (5): 625.

Unsurprisingly, a 2013 study by Nielsen, as part of its *Global Survey of Trust in Advertising*, found that earned media—also known as word-of-mouth media—is the most trusted source of information for consumers in all countries it surveyed worldwide and is the channel most likely to stimulate the consumer to action.[37] In marketing, the stimulation of the consumer to action is known as conversion. Indeed, this study indicated that 84% of consumers around the world say that they would trust word-of-mouth recommendations from friends and family—a form of earned media—above all other sources of advertising.[38] The second most trusted advertising source in 2013 was owned media, or owned advertising.[39]

For the purposes of this study, Nielsen defined earned media (or earned advertising) as advertising which is endorsed by a third party. Owned advertising was defined by Nielsen as advertising which is managed by a brand. Paid advertising was defined as advertising which is paid for by a brand.

For our purposes, the definitions, features and examples of these subcategories of digital media are provided in Table 1.2.[40]

Other authorities have made the distinction between online and offline earned media—or word-of-mouth advertising—and have been able to demonstrate that offline, in-person, word-of-mouth advertising can be substantially more effective than word-of-mouth advertising that is solely based online. Distinguishing type of digital content becomes slightly more complex when we add in the concept of online user-generated digital content.

1.3 Data Storage

There are two types of digital information: input and output data. Users provide the input data. Computers provide output data. A computer cannot compute anything or produce output data without the input of the computer user. With data storage space, users can save data onto a device. And should the computer power down, the data is retained. And instead of manually entering data into a computer, users can instruct the computer to pull data from storage devices. Computers can read input data from various sources as needed, and it can then create and save the output to the same sources or other storage locations. Users can also share data storage with others.

---

[37] Nielsen Holdings N.V. is a global information and measurement company with leading market positions in marketing and consumer information, television and other media measurement, online intelligence and mobile measurement.

[38] "Nielsen: Earned Advertising Remains Most Credible Among Consumers; Trust in Owned Advertising on the Rise". www.nielson.com

[39] Bonus tidbit: Other authorities make the distinction between online and offline earned media/word-of-mouth and have shown that offline word-of-mouth has been found to be more effective than online word-of-mouth

[40] Information for this table was sourced from a compilation of online resources provided by the Digital School of Marketing and the Agency 33 Public Relations Firm.

Table 1.2 Digital media marketing: earned, owned and paid

Media type	Earned media	Owned Media	Paid Media
Definition	Media activity, or other consumer communication, that: • Is related to a company or brand **and** • Is not directly generated, placed, purchased or managed by the company or its agents	Media activity, or other consumer communication, that: • Makes a direct connection between the company and brand **and** • Is generated, owned and managed by the company or its agents **and** • Is communicated in channels which are controlled by the company	Media activity, or other consumer communication, that: • Is related to a company or brand **and** • Is generated by the company or its agents **and** • Is communicated in channels outside of the company's control, in exchange for payment
Connection	Indirect connection to the company	Direct connection to the company	Direct, or indirect, connection to the company
Creation	Not created by the company	Created by the company	Created by, or for, the company
Control	No company control of content or communication	Direct company control of content and communication	Indirect company control of content and communication
Online examples	• Public and media relations • Traditional publicity mentions in professional digital media outlets • Viral marketing campaigns • Social media shares • Co-branding • Charity tie-ins • Influencer outreach • Public engagement • Consumer experience • Word-of-mouth/referrals	• Websites • Social media • Company blog content • Subscription newsletters • Sourced consumer reviews • Webinars • Videos/tutorials • Podcasts • Shareable content	• In-app advertising • Pay-per-click • Display/videos • Social media • Retargeting. • Sponsored posts/content • Sponsored tweets • Twitter cards • Fan acquisition • Lead generation • Brand ambassadors.
Offline examples	• News media coverage • Traditional publicity mentions in professional media outlets • Community building, development, service, engagement, partnerships • Charity tie-ins • Co-branding • Influencer outreach • Public engagement • Consumer experience • Word-of-mouth/referrals	• Traditional in-store visual merchandising and displays • Company press and media releases • Posters • Company mail-outs • Brochures	• Traditional advertising • Television commercials • Radio commercials • Print advertising • Outdoor marketing • Physical signage • Sponsorships • Direct mail

Table 1.3 Differences between NAS and SAN

Network-Attached Storage (NAS)	Storage Area Network (SAN)
• Single storage device or RAI • File storage system • TCP/IP ethernet network • Limited users • Limited speed • Limited expansion options • Lower cost and easy setup	• Network of multiple devices • Block storage system • Fibre channel network • Optimised for multiple users • Faster performance • Highly expandable • Higher cost and complex setup

1.3.1 Physical Storage Devices

Today, organisations and users require data storage to meet today's high-level computational needs like big data projects, artificial intelligence (AI), machine learning and the Internet of Things (IoT). And the other side of requiring huge data storage amounts is protecting against data loss due to disaster, failure or fraud. So, to avoid data loss, organisations can also employ data storage as backup solutions. To store data, regardless of form, users need storage devices.

1.3.1.1 Direct Data Storage

Direct area storage, also known as direct-attached storage (DAS), is as the name implies. This storage is often in the immediate area and directly connected to the computing machine accessing it. Often, it's the only machine connected to it. DAS can provide decent local backup services, too, but sharing is limited. DAS devices include floppy disks; optical disks—compact disks (CDs) and digital video disks (DVDs)—hard disk drives (HDD); flash drives; and solid-state drives (SSD).

1.3.1.2 Network-Based Data Storage

Network-based storage allows more than one computer to access it through a network, making it better for data sharing and collaboration. Its off-site storage capability also makes it better suited for backups and data protection. Two common network-based storage setups are network-attached storage (NAS) and storage area network (SAN).

NAS is often a single device made up of redundant storage containers or a redundant array of independent disks (RAID). SAN storage can be a network of multiple devices of various types, including SSD and flash storage, hybrid storage, hybrid cloud storage, backup software and appliances, and cloud storage. Table 1.3 explains some of the ways that NAS and SAN differ.

1.3.2 Cloud Storage

Cloud storage is a service model in which data is transmitted and stored on remote storage systems, where it is maintained, managed, backed up and made available to users over a network—most often via the Internet. Users of these services generally pay for their cloud data storage on either a per-consumption model or a standard monthly rate. Cloud storage includes data storage contained on the public cloud, in data centres, as well as in data mines.

Cloud computing is the delivery of on-demand computing resources (from applications to data centres) over the Internet on a pay-for-use basis. Cloud services can be deployed using three different models: public, private and hybrid.

In the public cloud model, cloud resources (i.e. hardware, software and infrastructure) are owned and operated by a third-party cloud service provider. Organisations consume IT services over the Internet from a shared pool of resources that are logically separated from one another. In the private cloud model, computing resources (that can be located on the premises or hosted by a third-party cloud service provider) are dedicated to one organisation and maintained on a private network. In the hybrid cloud model, both public and private clouds are used with data and applications communicating between the two.

The many different varieties of data storage options are featured in Fig. 1.5 below, including cloud-based and physical infrastructure, internal and external, public and private, small scale and commercial and hybrid options.

These data storage options are described with their features provided in Table 1.4, including the type of storage, the type of infrastructure, privacy features and scale of storage.

1.3.3 Cloud Infrastructure: Benefits and Challenges

Cloud computing is a fundamental shift in the delivery of information technology services, and governments must be prepared to leverage this alternative model of service delivery—and its benefits—as it evolves. That said, these benefits also present us with some unique challenges. This is because the nature of cloud computing results in an organisation or entity having a modified scope of control over its information (data) assets. This has resulted in a number of challenges related to data control, data protection and data privacy, some of which have already been noted in courts.

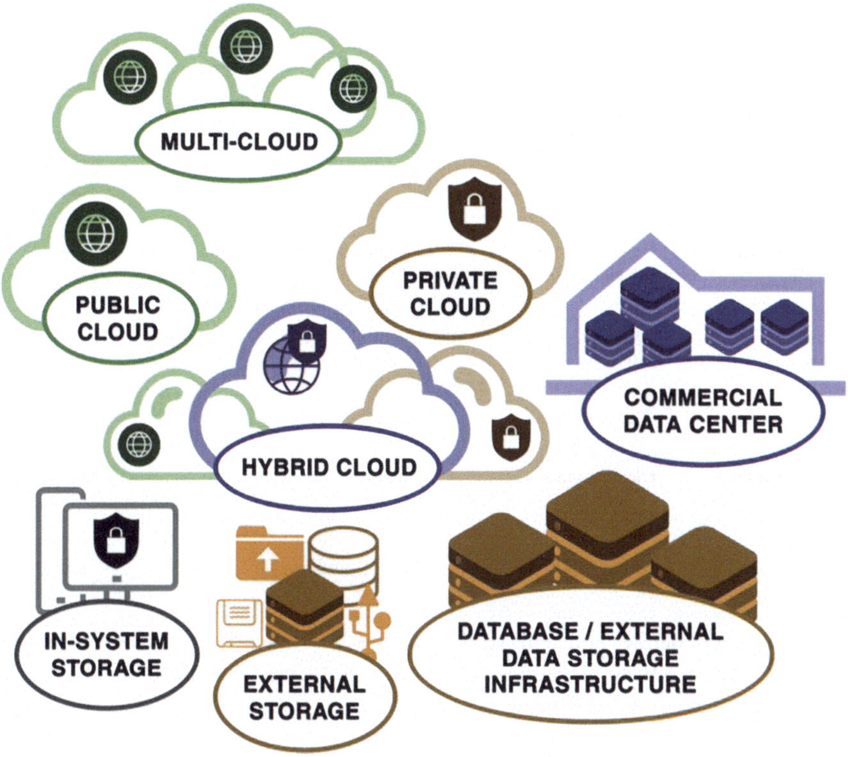

Fig. 1.5 Data storage solutions

1.3.3.1 Benefits

Some of the benefits of cloud data storage include renewed information technology infrastructure, closure of cyber hygiene gaps, availability of solutions and digital transformation.

Renewed Information Technology Infrastructure

The government's mission-critical information technology infrastructure is ageing and at risk of breaking down. It must be renewed. The transformation of information technology infrastructure is proceeding more slowly than anticipated, in part due to the complexity and challenges of consolidating the data centres, networks and email systems of 43 departments. For example, as stated in the *Shared Services Canada Resource Alignment Review*, cloud computing presents a significant opportunity to

Table 1.4 Features of data storage options

Storage type	Infrastructure	Privacy	Scale
Direct data storage	Physical hardware directly connected to the machine	Localised to holder	Small
Network-based data storage	Physical hardware connected through network	Localised to holder	Small/ medium
Database/external data storage infrastructure	External physical infrastructure	Localised to infrastructure operators and subscribing individuals/ organisations	Large
Commercial data centre	External physical infrastructure	Localised to infrastructure operators and subscribing individuals/ organisations	Large
Public cloud	Cloud-based	Public	Medium/ large
Private cloud	Cloud-based	Private	Small/ medium
Hybrid cloud	Cloud-based	Mix of both public and private	Large
Multi-cloud	Cloud-based	Mix of public	Large

address these challenges and achieve cost savings.[41] A key driver for cloud adoption is the need to move from managing ageing assets to focusing on using commercially available services to deliver business value.

Closure of Cyber Hygiene Gaps

The rising number of security incidents has demonstrated the inconsistent security posture of government systems. Lessons learned have highlighted challenges in the government's ability to quickly identify assets and perform timely patching and remediation of known vulnerabilities. These challenges leave the government exposed to cyber threats. Moving to the cloud provides the government with an opportunity to deploy secure virtual environments rapidly and also to change its approach to managing its assets and addressing security from the outset.

Cloud service providers have a significant budget to maintain, patch and secure their cloud infrastructure. This means that cloud service providers can mitigate many common risks that government organisations often face while also providing a much stronger overall security posture. Leading cloud service providers also subject their services and processes to multiple third-party audits and meet numerous

[41] *Shared Services Canada Resource Alignment Review* (2016). Available online at: https://www.canada.ca/en/treasury-board-secretariat/corporate/reports/shared-services-canada-resource-alignment-review.html

internationally recognised industry certifications to provide valuable assurance to their clients.

Availability of Solutions

Increasingly, the industry is providing only public cloud solutions or focusing their efforts on cloud services. The expression "born in the cloud" signifies that a particular solution has been designed to reside natively in a public cloud and that no software equivalent exists. Excluding cloud solutions from the government's information technology landscape would limit the solutions that are available.

Digital Transformation

The government's demand for information technology capacity and capabilities exceeds the available supply. The government wants to follow the global trend of improving digital services for its citizens, and this can be greatly enabled by cloud services. Many governments have put into effect "cloud-first" policies, including in the other countries that make up the Five Eyes: the United States, the United Kingdom, Australia and New Zealand. Failing to make the cloud part of the government's digital landscape will impact how quickly the government's digital vision can be achieved.

1.3.3.2 Challenges

The main challenges in cloud data storage relate to data security, data residency and data sovereignty.

Data Security

When data is moved to the cloud, security becomes a shared responsibility with cloud service providers. Cloud service providers are responsible for the security of the cloud, and government departments are responsible for security to the cloud, from the cloud and in the cloud. A simple analogy could be the example of a rental building. Property owners or managers are responsible for the security of the building and common areas, but renters are responsible for the security measures within the individual units that they rent. The modified scope of control, combined with the shared responsibility, often leads to fears that the security posture of an organisation may be degraded through this model.

Data Residency

Cloud service providers are typically hyper-scale providers, meaning that they have global deployment on a scale that eclipses the government's own information technology assets many times over. Although cloud service providers have the ability to move a client's data anywhere in the world, the client typically has the option to isolate their data to a given geographical region. This is an important feature because it allows the client to keep and maintain data under the laws and policies of a particular territory. For example, moving the government's data outside of the geographic boundaries of the nation over which it governs could impact the government's access to data and services that are vital to its business continuity.

Data Sovereignty

Regardless of where the cloud resources are physically located, when data is stored in a cloud environment, the stored data may be subject to the laws of other countries. As previously mentioned, cloud service providers are hyper-scale providers that have global deployment. A cloud service provider with foreign operations could be required to comply with a warrant, court order or subpoena request from a foreign law enforcement agency seeking to obtain government data. This means that countries like Canada cannot ensure full sovereignty over its data when it stores data in the cloud. Lack of full data sovereignty has the potential to damage the government and third parties. Sensitive government data could be subject to foreign laws and be disclosed to another government. Under some foreign laws, disclosure of government data could take place without notice to the government.

Table 1.5 summarises the various data storage methods that we have discussed in this section.

1.4 Data Sovereignty

Sovereignty refers to the legal power and authority of a governing body to exercise control over a nation or state and the autonomy of that governing body to do so freely and without external pressure, control or influence.[42]

In any nation or state, sovereignty is assigned to the person, governing body or other institution that has the ultimate authority over other people in order to establish a law or change an existing law.[43] Nations and states are also sometimes described as

[42] Philpott, D. (2020). Sovereignty. In Zalta, E. N. (ed.), The stanford encyclopedia of philosophy (Fall 2020 ed.), Metaphysics Research Lab, Stanford University.

[43] Spruyt, H. (1994). *The sovereign state and its competitors: An analysis of systems change* (pp. 3–7). Princeton University Press.

Table 1.5 Data storage methods

Storage method	Concept	Benefits	Challenges
Direct data storage	• Owned and operated either in-house	• Easy to access • Directly managed by user	• Limited size • Security
Network-based data storage	• Owned and operated either in-house or through a third-party provider • Offered to selected users over the network	• Central file storage and access • Easy to set up and use • Reduce hardware costs • Space-saving and compact	• Infrastructure • Security • Corruption • Scale
Database storage	Owned and operated either in-house or through a third-party provider • Offered to selected users over the network	• Reduced data redundancy • Reduced updating errors and increased consistency • Greater data integrity and independence from application programs • Improved data access to users through use of host and query languages • Improved data security • Reduced data entry, storage and retrieval costs	• Security • Compatibility. • Cost. • Corruption.
Commercial data Centre	• Provided by service providers and shared among multiple users • Available to anyone over the public internet.	• Protecting power • Cutting costs • Improving efficiency	• Requires significant infrastructure to support the centre's hardware and software include power subsystems, uninterruptible power supplies (UPS), ventilation, cooling systems, etc. • Needs network security appliances such as firewall and intrusion protection • Needs application delivery assurance
Public cloud	• Provided by third-party cloud service providers and shared among multiple tenants • Available to anyone over the public Internet • Storage capacity can be scaled quickly and conveniently, as the demand for storage increases	• Low starting costs and fees over time • Little to no maintenance, as it's carried out by the provider • No need to provide for hardware resources, as these are provided by the cloud provider • Large choice of services offered by your provider,	• Limited flexibility to meet specific/unique business needs • Added security is always recommended • Enhanced encryption and security methods should still be used on top of public cloud security features • User subscription fee is

(continued)

Table 1.5 (continued)

Storage method	Concept	Benefits	Challenges
		which means less time spent installing and configuring • Can often pay for only the services used and the amount of usage • Unlimited or almost unlimited scalability to increase storage capacity as you need • Initial access to a small amount of storage is free • Responsibility for securing services and data is shared between the provider and the user • Sensitive data can be protected through services offered by the provider	often required to unlock access to larger storage capacities • Sensitive data is often stored elsewhere, the location of which may be unknown to the user • Cloud infrastructure services may differ from one provider to another, creating a vendor lock-in
Private cloud	• Owned and operated either in-house or through a third-party provider • Offered to selected users over the Internet or through a private internal network • Hosted on private servers	• Can start small and increase infrastructure as needed • Apps and data can be contained in-house, giving greater control to the organisation • Provides users with greater control over security and functionality • Potential for improved security protocols to be added, as needed. • Low maintenance costs • Customizable functionality, allowing for storage to be business-specific • Highly scalable—easy to add or remove storage capacity, as needed	• Requires traditional data centre staffing and maintenance • Potential for high initial startup costs • Must provide hardware resources, installation, configuration and management of the cloud infrastructure itself, aside from apps and data • If there is a shortage or a lack of access to enough hardware resources, it can be difficult to provide for geo-diversification and redundancy of locations • Less ability to prevent major service outages if there is a problem with your primary location
Hybrid cloud	• Combination of both public and private cloud • Shared security responsibility	• Creates a more varied cloud environment • Helps maintain tighter controls over sensitive data processes	• Lack of readiness for data migration • Poorly defined SLAs (service-level agreements) • Governance problems • Compatibility issues

(continued)

Table 1.5 (continued)

Storage method	Concept	Benefits	Challenges
			• Lack of visibility and control
Multi-cloud	• Use of a mixture of different public clouds to host data, rather than combining the public and private clouds • Can also involve the use of private clouds, but not necessary in this model	• Can cover a wide variety of business needs that cannot be provided by one single cloud provider • Cost-effective method of ticking off all your boxes for cloud needs • Eliminates the feeling of dependency on one cloud provider • High scalability with the ability to branch out to different providers if new IT needs emerge	• Increased management complexity • Cost control • Governance, compliance and security concerns • Interoperability

being sovereign themselves. When a nation identifies as a sovereign nation, it means that the population or residents of the nation have power over themselves and control over their own government, rather than their government being under the control of an external authority. In this way, sovereignty in domestic law refers to the power of a governmental body to rule without intrusion from outside forces or external influence from other countries.

In international law, sovereignty is the exercise of power by a state.[44] De jure sovereignty refers to the legal right of the state to exercise their power and authority. This can be contrasted with de facto sovereignty, which refers to the actual factual ability of a state to exercise their power and authority rather than the legal right to do so. While we typically expect de jure and de facto sovereignty to exist at the same place, at the same time and within the same governing organisation, a failure of that expectation can become an issue of concern for the determination of sovereignty in international law.

1.4.1 Sovereignty for Data

Data sovereignty differs from data residency in that not only is the data stored in a designated location but is also subject to the laws of the country in which it is physically stored. This difference is crucial, as data subjects (any person whose personal data is being collected, held or processed) will have different privacy and

[44]Wallerstein, I. (2004). *World-systems Analysis: An Introduction* (p. 44). Duke University Press.

security protections according to where the data centres housing their data physically sit.

This difference is also crucial for businesses, as a government's rights of access to data found within its borders differ widely from country to country. This is where data sovereignty and residency are often conflated. Ensuring data sits within a geographical location for whatever reason—whether avoiding or taking advantage of laws, regulations and tax regimes or even for pure preference and comfort—is a matter of data residency. But the principle that the data is subject to the legal protections and punishments of that country is a matter of data sovereignty.

They are clearly related, and even two sides of the same coin, but one is a matter of national legal rights and obligations, while the other is a matter of geography. Recognising this distinction will help professionals better prepare for compliant data management and exchange.

1.4.1.1 Example: Canada

We will cover a variety of different national approaches to data residency, localisation and sovereignty in Chap. 5. With respect to Canada, "data sovereignty" is Canada's right to control access to and disclosure of its digital information subject only to Canadian laws. The Government of Canada, in their White Paper on *Data Sovereignty and Public Cloud*, outlines the governmental strategy whereby cloud services are identified and evaluated as the principal delivery option when initiating information technology (IT) investments, initiatives, strategies and projects.[45]

1.4.2 Governmental Risks

In the public cloud environment, government data is entrusted to a third party that may be subject to the laws of a foreign country, even if the data resides in another. As such, the key risk to the government with respect to data sovereignty is that foreign agencies can leverage laws in their home country to compel cloud service providers to turn over the government's data. This could directly damage the governmental authority or cause citizens to lose trust in their government.

1.4.2.1 Example: United States and Canada

It should be noted that many countries, including Canada, have laws that allow them to subpoena private organisations or obtain a warrant for information from such

[45] https://www.canada.ca/en/government/system/digital-government/digital-government-innovations/cloud-services/gc-white-paper-data-sovereignty-public-cloud.html

organisations to support legal investigations. For example, the primary risk to data sovereignty is the US's *Foreign Intelligence Surveillance Act* (FISA) and the Government of the United States' ability to compel an organisation subject to the US law to turn over data under its control, regardless of the data's location and without notifying Canada. It is important to note that there are long-standing information-sharing agreements between security and law enforcement agencies in Canada and the United States. Those long-standing agreements, as well as mutual legal assistance treaties, are other methods for the United States to obtain access to information held in Canada.

1.4.3 Risk Mitigation Strategies

There are a number of mitigations that the government is recommending reducing the risk of unauthorised access to sensitive information. They include limiting the categories of data stored in the cloud, encrypting the data and using standard contract clauses.

1.4.3.1 Limiting Categories

Following the lead of other countries, it is important to limit the categories of data that may be stored in the cloud. To date, major cloud service providers have asserted their ability to meet the government's requirements for storing and processing data. This assertion will be tested by the government as it procures cloud services for protected scenarios. For this reason, the government has stated that only data up to and including protected data may be deployed to a public cloud.

1.4.3.2 Encrypting Data

When using cloud services, encryption can provide a high degree of segregation to protect personal and other sensitive information. Data that is encrypted using a strong cryptographic algorithm is protected from anyone who does not have the decryption key. The government will be directing that all protected data hosted in the cloud be appropriately encrypted while in transit and at rest and that the government maintain exclusive control of the encryption keys. It should be noted that if the encrypted data is to be processed in the cloud by an application, the data would need to be decrypted before processing. As a result, an unencrypted snapshot of the data would be stored temporarily in the cloud before being overwritten. The government will continue to investigate techniques available to mitigate this risk and prevent the exposure of data to unauthorised parties. Possible mitigation techniques include data anonymisation and hardware-level isolation.

1.4.3.3 Using Standard Contract Clauses

Contracting authorities should ensure that all contracts with cloud service providers include clauses that compel the cloud service provider to disclose all unauthorised access to data, including access made under court order, where applicable and unless the cloud service provider is prohibited from doing so by law. There are some cases where such disclosure is forbidden under federal law, such as in the case of an order by the US Foreign Intelligence Surveillance Court. The government may request that cloud service providers provide their procedures for navigating the conflict between their contractual obligations and applicable laws.

1.5 Residency and Localisation

The terms "**data sovereignty**", "**data residency**" and "**data localisation**" are often a source of confusion for businesses managing data across borders, especially on cloud infrastructure. They are in fact used so interchangeably that their individual meaning has become lost. This interchangeable use is portrayed most clearly by the remarkable consistency in the number of searches each term receives in search engines like Google.

Residency, sovereignty and localisation—with respect to data—are effectively three degrees of a single concept: how data privacy impacts cross-border data flows. This subject has become increasingly important in recent years, as evidenced, for example, by the *General Data Protection Regulation* (GDPR) legislation granting greater rights to individuals over how their data is used by businesses both within and outside the European Union. Described in the simplest terms:

Data sovereignty refers to the jurisdictional control or legal authority that can be asserted over data because its physical location is within jurisdictional boundaries. It is the right of a state to control access to and disclose its digital information subject only to the laws of that state.

Data residency refers to the physical or geographical location of where the data of an organisation or entity is stored. This is an important determiner for commercial legal purposes as well as taxation.

Data localisation refers to various policy measures that restrict data flows by limiting the physical storage and processing of data within a given jurisdiction's boundaries. It can also refer to restrictions placed on the ability of companies to move, store, process or otherwise handle their users' personal data. Multiple countries have adopted localisation policies to combat multiple concerns over the free flow of data.

Data localisation/residency laws require data about a nation's citizens or residents to be collected, processed and/or stored inside the country, often before being transferred internationally.

1.5.1 Data Residency

Data residency refers to where a business, industry body or government specifies that their data is stored in a geographical location of their choice, usually for regulatory or policy reasons.

A typical example of a data residency requirement in action is where a company wishes to take advantage of a better tax regime. Doing so will usually require the business to prove they are not conducting too great a proportion of core business activities outside that country's borders—including the processing of data. They will therefore impose a data residency that requires them to use certain infrastructures and then impose strict data management workflows on themselves and any cloud service providers in order to protect their taxation rights.

1.5.1.1 Data Residency-as-a-Service

Data residency is the localisation of regulated data such as personal information within a particular region or country. That could include only storing the data, but it could also include processing it. Where this data is processed in accordance with the laws of that specific region. As we become more and more connected around the globe, we can see regional conflicts arise. For example, what may be seen as an acceptable use of personal information in Egypt could be controversial in China.

Companies which provide data residency-as-a-service facilitate the tailoring of data processing, through internal processing decisions, and depending on regional expectations. These types of organisations, those that handle international data, must ensure that data privacy is not put at risk when data is shared across borders. For this reason, it is necessary for these companies to have a full understanding of the legal requirements of storing data in a certain country. This understanding is fundamental to meeting data privacy and security standards.

1.5.2 Data Localisation

Data localisation requires companies that use, store or process data online to do so in the country of its origin. The main drivers behind this trend include:

- Information security
- Concerns over data privacy
- National data sovereignty
- Law enforcement
- Potential support for local businesses

Data localisation, like data sovereignty, is a version of data residency predicated on legal obligations. It is the most stringent and restrictive of these terms, but it is also the concept that is growing the fastest internationally.

Data localisation laws are regulations used to restrict data movement across borders and pose certain limitations to the global digital economy, as well as to the ability of nations to maximise economic and social benefits of data-reliant technologies and businesses. Such regulations heavily depend on the country. More and more countries are implementing or introducing long-term regulations, which will control the free flow of data.

Data localisation in law requires that data created within certain borders stay within them. In contrast to the two terms above, it is almost always applied to the creation and storage of personal data, with exceptions including some countries' regulations over tax, accounting and gambling. In many cases, data localisation laws simply require that a copy of such data be held within the borders of a country, usually to guarantee that the relevant government can audit data on its own citizens, provided that there is due cause, without having to contend with another government's privacy laws. India's draft *Personal Data Protection Bill* is an example of this. However, there are countries where the law is so strict as to prevent it from crossing the border at all. For instance, Russia's *On Personal Data Law* (OPD Law) requires the storage, update and retrieval of data on its citizens to be limited to data centre resources within the Russian Federation.

Typical criticism of these laws is that they use the mask of enhanced cybersecurity or citizens' privacy concerns to conceal the real motivation of national protectionism. This border-based siloing of data, it is claimed, obstructs businesses and governments from realising the full potential that data stands to offer and contributes to digital factionism and the "splinternet". But regardless of the debates surrounding the practices, corporations and organisations often need to better understand the importance of the differences between these three terms. We may be sticklers for using the right vocabulary in the correct circumstances, but the frequency and manner in which these words are used interchangeably among the businesses we engage with, and even other industry commentators, indicates a widespread misunderstanding.

Any company seeking international growth will sooner or later face the question of whether localising software products is required or not. There is no simple answer to this question, as it all depends on a multitude of aspects related to the business and its specific needs.

Companies dealing with complaint data need to create and maintain multiple data centres and complex infrastructures in different jurisdictions, which adds extra cost to operational spending. As well, companies relying upon such services may find they avoid certain markets altogether due to the increased cost of doing business there. This then further affects the attractiveness of different regions when it comes to capital investment and talent retention, with data localiwation restrictions acting as digital walls between countries. Yet, there are clear indicators and proof points to determine the business value of localisation and prepare a solid business localisation strategy and localisation checklist. As new cybersecurity and data laws get more and more power, it is important to understand the potential repercussions of avoiding them.

Operating in countries with such laws requires that companies remain compliant to local regulations, and this, of course, requires additional costs. Companies should either comply and invest in infrastructure and human capital to provide services or use third-party services that will help them stay compliant. This also saves time, resources and costs. For companies that do not comply, there is a risk of getting fined and being legally prosecuted by local authorities.

Another issue that might arise is when a copy of the data must be maintained in the country where the service is being provided. This can introduce additional costs and infrastructure risks as adding, deleting or updating records must be done across multiple locations. If not implemented properly, this can also risk having inconsistent datasets across countries as updates or deletion of some records might happen in some, and not all, countries where data is stored. As specific restrictions on cross-border and interjurisdictional data transfers among global economies become more strict, this results in additional costs and preparation time for companies seeking to scale globally.

1.5.3 Localisation Models

There are four primary types of data localisation legal frameworks:

1. Total data sovereignty
2. Select data sovereignty
3. Data replication
4. Controlled localisation

1.5.3.1 Total Data Sovereignty

These are the countries that require that all data related to their citizens be stored in servers physically located inside the country. This applies to data use in all industries.

Example Passed in August 2021, the *Personal Information Protection Law* of the People's Republic of China took effect on November 1, spelling out rules around data collection, use and storage, as well as what international companies must do when they transfer data out of the country.

1.5.3.2 Select Data Sovereignty

This strategy is based on the premise that only data which is related to certain industries cannot leave the country borders (usually such industries are financial services, healthcare, telecommunications and government/defence). This applies to data use in select industries only.

Example In 2018, the Reserve Bank of India declared that all payment system providers should store payment data in the country, with similar measures planned for e-commerce, social media, telecom and healthcare. The UAE and Australia have similar measures for healthcare data, while Turkey introduced data localisation for financial services.

1.5.3.3 Data Replication

In this model a copy of all compliant data needs to be stored locally and then can cross the country borders.

Example The revised *Indian Personal Data Protection Bill* (PDPB)[46]—like the EU's *General Data Protection Regulation*[47]—requires tech firms to have consumers' consent before collecting and processing their data. The data mirroring requirement of this law requires that a copy of data on Indian citizens be stored in India. Similarly, the new *Cybersecurity Law of 2019* passed by the Vietnamese National Assembly requires online service providers to store citizens' personal data inside the country.

1.5.3.4 Controlled Localisation

This model is based on having less extreme laws that focus on data privacy.

Example The most obvious case is the EU's *General Data Protection Regulation* (GDPR),[48] which allows data transfers to other countries under certain conditions. Data can be transferred to countries that have the same level of data protection as in

---

[46] "The Personal Data Protection Bill, 2019" (PDP Bill 2019) *PRSIndia*. 11 December 2019. http://prsindia.org/billtrack/personal-data-protection-bill-2019

[47] *EU General Data Protection Regulation (GDPR):* Regulation (EU) 2016/679 of the European Parliament and of the Council of 27 April 2016 on the protection of natural persons with regard to the processing of personal data and on the free movement of such data, and repealing Directive 95/46/EC (General Data Protection Regulation), OJ 2016 L 119/1.

[48] *EU General Data Protection Regulation (GDPR):* Regulation (EU) 2016/679 of the European Parliament and of the Council of 27 April 2016 on the protection of natural persons with regard to the processing of personal data and on the free movement of such data, and repealing Directive 95/46/EC (General Data Protection Regulation), OJ 2016 L 119/1.

the European Union. Similar laws have been passed in Brazil—the *General Data Protection Law of 2018* (LGPD)[49,50]—as well as in Colombia and Peru.

These localisation models are provided with their definitions and examples in Table 1.6.

1.6 Legal Considerations

However, there is an important distinction between these terms. Regardless of where the cloud infrastructure and resources are physically located, data stored in a cloud environment may end up being subject to the laws of other countries.

Recognition of this distinction can be useful in helping professionals to better prepare for compliant data management, consumer data privacy and data exchange obligations.

Consider, for example, if the physical storage location of governmental data was moved outside of the geographical boundaries of the nation over which it has governing control. Such a move could have a huge impact on that government's access to data and services that are vital to its business continuity. Similarly, for corporations who need access to cloud storage, data residency and data sovereignty are important features to consider because they allow the corporation—as a client of the cloud service provider—to keep and maintain data under the laws and policies of a particular governing territory.

Rehashed in the simplest terms: data residency refers to the geographic location of the data storage infrastructure or cloud storage resources which store the data; data sovereignty refers to the laws and policies under which the data is governed; and, finally, data localisation refers to the set of requirements provided by geographic data residency laws, as established by a country, in order for that country to maintain sovereignty over its own data.

These concepts are crucial for corporations to consider, as the implications of where their data storage facilities are located have a wide scope of impact on the corporate social responsibility, the laws which apply to them and their range of influence. Storage options range from physical or on-site storage to cloud-based and off-site third-party storage.

Figure 1.6 below illustrates the data storage options which are available in the digital world.

[49]The *Lei Geral de Proteção de Dados* (LGPD) is a legal framework to regulate the collection and use of personal data which came into effect in Brazil on August 16, 2020. The law was passed, and will be enforced, by the Autoridade Nacional de Proteção de Dados or National Data Protection Authority (ANPD).

[50]*Lei Geral de Proteção de Dados*. 2018. Available online and in English at: https://iapp.org/resources/article/brazilian-data-protection-law-lgpd-english-translation/

Table 1.6 Localisation models

Localisation model	Definition	Examples
Total data sovereignty	• These are the countries that require that all data related to their citizens be stored in servers physically located inside the country • This applies to data use in all industries	• Passed in August 2021, the *Personal Information Protection Law* of the People's Republic of China took effect on November 1, spelling out rules around data collection, use and storage, as well as what international companies must do when they transfer data out of the country
Select data sovereignty	• This strategy is based on the premise that only data which is related to certain industries cannot leave the country borders • Usually these industries are financial services, healthcare, telecommunications and government and/or defence • This applies to data use in select industries only	• In 2018 the Reserve Bank of India declared that all payment system providers should store payment data in the country, with similar measures planned for e-commerce, social media, telecom and healthcare • The UAE and Australia have similar measures for healthcare data, while Turkey introduced data localisation for financial services
Data replication	• In this model a copy of all compliant data needs to be stored locally and then can cross the country borders	• The revised *Indian Personal Data Protection Bill* (like the EU's GDPR) requires tech firms to have consumers' consent before collecting and processing their data • The data mirroring requirement of this law requires that a copy of data on Indian citizens be stored in India • Similarly, the new cybersecurity law of 2019 in Vietnam requires online service providers to store citizens' personal data inside the country
Controlled localisation	• This model is based on having less extreme laws that focus on data privacy	• The most obvious case is the EU's *General Data Protection Regulation* (GDPR), which allows data transfers to other countries under certain conditions • Data can be transferred to countries that have the same level of data protection as in the EU • Similar laws have been passed in Brazil (*General Data Protection Law of 2018*), Colombia and Peru

Those data storage options which are provided above the broken line are those options which are cloud-based in nature. Those below the broken line are those storage options which are physically based in nature.

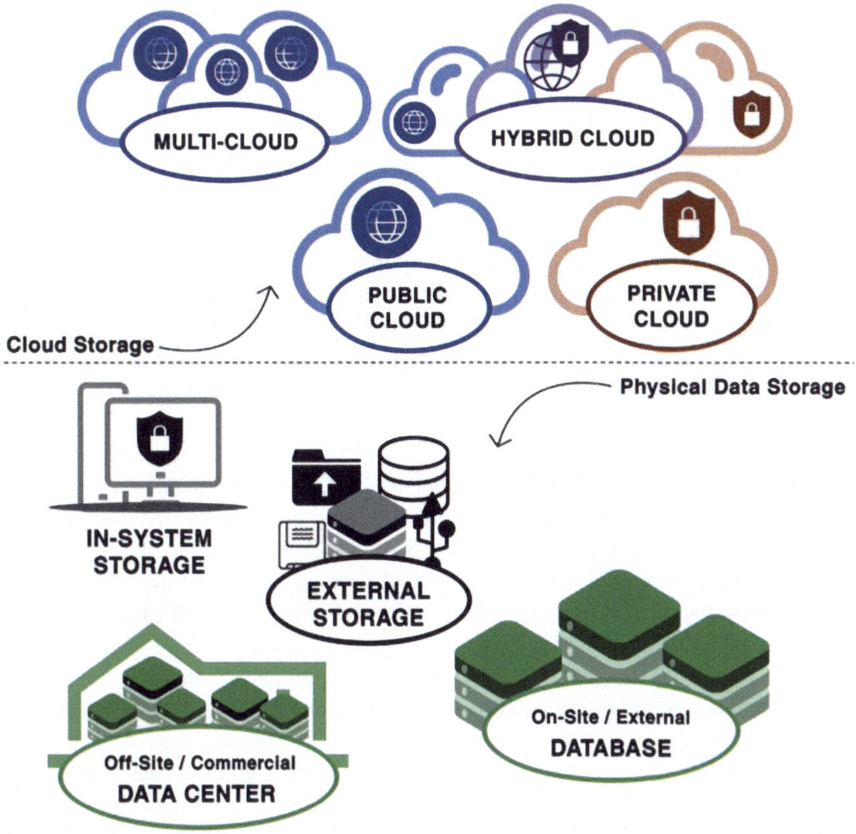

Fig. 1.6 Data storage options

1.7 Summary

The increased digitisation of organisations, driven by the rapid adoption of technologies such as cloud computing and data analytics, has increased the importance of data dramatically, influencing traditional industries as well as the digital economy. Data is rightfully considered the lifeblood of the modern global economy, while transferring protected data overseas gets more and more complicated. More and more countries develop barriers that make this process time-consuming and costly due to newly enacted data residency regulations. Data residency is the localisation of regulated data such as personal information within a particular region or country. That could include only storing the data, but it could also include processing it. Where this data is processed in accordance with the laws of that specific region.

As information age progresses, geographies' relevance to privacy is becoming more and more important. While large, multinational privacy regulations like the

General Data Protection Regulation (GDPR)[51] or major laws like the *California Consumer Privacy Act* (CCPA)[52] make headlines, there are countless smaller, regional laws and customs that often get less coverage or attention. These laws are often the cornerstone obstacle in the global expansion plans for the multinational businesses.

Despite the significant benefits to companies, consumers and national economies that arise from the new digitalised world and ability of organisations to easily share data across borders, dozens of countries have erected barriers to cross-border data flows. Among them is data residency requirements that confine data within a country's borders, or a concept also known as "data localisation".

Data localisation can be explicitly required by law or is the result of other restrictive policies that make it infeasible to transfer data. They require companies to store a copy of the data locally, process data locally and mandate individual or government consent for data transfers.

Data sovereignty can be a complex topic—legally, technically and politically—but it is not insurmountable. In this chapter, we have explored the concepts of data, first through the lens of ownership and property law, then in terms of data privacy and security. We discussed the fundamentals of data storage and the challenges that can present when dealing with transnational storage infrastructure. Finally, we outlined the concepts of data residency and data localisation both in contrast and in tandem with data sovereignty.

In the next chapter, we will explore the concept and realisation of digital governance and its intersection with data sovereignty on both a national and an international level.

The answers to the following questions are provided within this chapter:

1. What is data security?
2. What are three examples of personal data?
3. How do data residency, data sovereignty and data localisation differ?
4. Why might a country care about residency, localisation and sovereignty with respect to data?
5. Why might a corporation care about residency, localisation and sovereignty with respect to data?

[51] *EU General Data Protection Regulation (GDPR):* Regulation (EU) 2016/679 of the European Parliament and of the Council of 27 April 2016 on the protection of natural persons with regard to the processing of personal data and on the free movement of such data and repealing Directive 95/46/EC (General Data Protection Regulation), OJ 2016 L 119/1.

[52] *California Consumer Privacy Act.* Section 3, Title 1.81. 5 of the CCPA, added to Part 4 of Division 3 of the California Civil Code.

Chapter 2
Digital Governance

Governance has been defined to refer to structures and processes that are designed to ensure accountability, transparency, responsiveness, rule of law, stability, equity and inclusiveness, empowerment, and broad-based participation. With respect to digital governance, it refers to the structures and processes which are designed to ensure accountability, transparency, responsiveness, rule of law, stability, equity and inclusiveness, empowerment, and broad-based participation, of all varieties, in the online world, rather than that which is limited to the physical world.

Digital tools and social media have empowered people through widespread access to information and global connections. Citizens are using technology to hold governments to account and to exercise their civic rights. Governments are using technology to be more transparent, accountable and inclusive. They are also innovating solutions to pressing issues, including delivering services better, establishing cutting edge systems of procurement, advancing fiscal openness and fighting fraud and abuse.

Yet, the same technologies can represent real risks to democracies. Unaccountable institutions are leveraging technology to pursue their own interests in and across markets with little to no—or conflicting—jurisdiction and accountability. And public institutions are dealing with the unintended consequences of fast-moving technologies that often outpace legal safeguards and government oversight. Some of the recommendations that have been put forward with regard to digital governance involve inclusive digital transformation and innovation, digital rights-based frameworks, and safeguards to prevent misuse of digital technologies.

M. Lukings, A. Habibi Lashkari, *Understanding Cybersecurity Law in Data Sovereignty and Digital Governance*, Progress in IS,
https://doi.org/10.1007/978-3-031-14264-2_2

2.1 Topic Overview

In this chapter, we will explore the concept of digital governance. We will begin by defining governance and some of the philosophies and theories of governance. These will be applied to nation/state governments as well as to corporations and organisations. We will outline the features of social responsibility, including sustainability, accountability, transparency, and the social contract. Having established an understanding of governance, we will proceed to look at governance through the digital framework, focussing on issues of corporate digital governance and the responsibilities which flow from that. Finally, we will discuss the growing issue of digital governance of user generated digital content, an issue which has been at the forefront of digital policy in many governments as of late.

2.2 Governance

Governance is the way rules, norms and actions are structured, sustained, regulated and held accountable. The degree of formality depends on the internal rules of a given organisation and, externally, with its business partners. Governance, as a whole, consists of all of the processes of interactions be they through the laws, norms, power or language of an organised society over a social system. It is done by the government of a state, by a market, or by a network.

2.2.1 Types of Governance

For our purposes, there are two main types of governance: good governance and bad governance. Strategic governance is the concept which applies to the interplay of governance and strategic decision-making.

2.2.1.1 Good Governance

Good governance is an approach to a type of government that is committed to creating a system founded in justice and peace that protects individual's human rights and civil liberties. According to the United Nations, "good governance" is measured by the eight factors of participation, rule of law, transparency, responsiveness, consensus oriented, equity and inclusiveness, effectiveness and efficiency and accountability.

1. **Participation—**requires that all groups, particularly those most vulnerable, have direct or representative access to the systems of government. This manifests as a strong civil society and citizens with the freedom of association and expression.

2. **Rule of Law**—is exemplified by impartial legal systems that protect the human rights and civil liberties of all citizens, particularly minorities. This is indicated by an independent judicial branch and a police force free from corruption.
3. **Transparency**—means that citizens understand and have access to the means and manner in which decisions are made, especially if they are directly affected by such decisions. This information must be provided in an understandable and accessible format, typically translated through the media.
4. **Responsiveness**—imply involves that institutions respond to their stakeholders within a reasonable time frame.
5. **Consensus Oriented**—is demonstrated by an agenda that seeks to mediate between the many different needs, perspectives and expectations of a diverse citizenry. Decisions need to be made in a manner that reflects a deep understanding of the historical, cultural and social context of the community.
6. **Equity and Inclusiveness**—depends on ensuring that all the members of a community feel included and empowered to improve or maintain their well-being, especially those individuals and groups that are the most vulnerable.
7. **Effectiveness and Efficiency**—is developed through the sustainable use of resources to meet the needs of a society. Sustainability refers to both ensuring social investments carry through and natural resources are maintained for future generations.
8. **Accountability**—refers to institutions being ultimately accountable to the people and one another. This includes government agencies, civil society and the private sector all being accountable to one another as well.

2.2.1.2 Bad Governance

The World Bank governance indicators are:

- Voice and accountability
- Political stability and absence of violence
- Government effectiveness
- Regulatory quality
- Rule of law
- Control of corruption

Based on the World Bank's governance indicators, the key causes for bad governance are twenty.

Lack of Voice and Weak Accountability

Governing bodies refusing to listen to the voice of those they govern and refusing to take accountability for their actions lead to bad governance. By ignoring the voice of those being governed, their opinions are no longer heard or taken into consideration by the governing body. Democratic governments focus on accountability as a method to ensure the public understands what's happening and provides them a way to proceed when things go wrong. Weak accountability in turn causes a distrust between the two parties and can lead to instability. This distrust and uncertainty create an unfavourable relationship between the parties.

Political Interference

Bad governance occurs as a consequence of frequent changes in government or "political instability". Instability in political regimes, such as a democracy, has been proven to coincide with poor governance.

Corruption

Bad governance is often considered to come hand in hand with corruption. Corruption occurs in many sectors ranging from political to economic environments. Corruption can occur in many different ways and forms. The existence of corruption within a governing body causes bad governance as the officials place their personal gains over others.

The impacts and consequences of bad governance are widespread and don't only affect the settings in which they occur:

Poor Economic Growth

Bad governance heavily impacts the per capita growth of a country. African countries have experienced this impact the most since World War II. The economic growth of a country is significantly impacted when exposed to indicators of bad governance, but difference indicators influence the degree of impact. A lack in regulatory quality, government's ineffectiveness and a lack of control on corruption have been linked to poor economic growth.

Failed State

When the public administration cannot provide many public services and the citizens do not recognise that their government is not legitimate, then the state would become a failed state.

Corruption

Corruption not only is a cause of but can also occur as a consequence of bad governance. There was a distinct link suggesting that higher levels of governance and a better environment to conduct business are impacted by the presence of corruption within an economy.

Solution: Transparency

The World Bank Institute has suggested that through introducing reforms that address transparency, the degree of bad governance can be levitated, and good governance can slowly take precedence. These reforms include:

- Disclosure to the public of all assets and income of candidates running for public office.
- Disclosure to the public of all political contributions to parties along with all documentation related to legislation and voting in parliament.
- Clear separation of laws relating to areas of governance, e.g. business, politics, etc.
- Release publicly any company that has been blacklisted due to bribery.
- Ensure access by the public to all government information.
- Ensure media has a freedom of speech.
- Implementation of International Monetary Fund's (IMF's) reports on standards and codes framework.
- Release of government reports on payment from extraction industries along with open meetings between two entities if a country's citizens are involved.
- Domestic bank's owners and financial reports are opened to the public.
- Competitive procurement (over the Internet) by governing entities is available to public.
- Introduction of governance and anti-corruption surveys supported by World Bank.
- Budgets and meetings at the city level are transparent.

Solution: Dealing with Corruption

A strong correlation exists between bad governance and corruption. By curbing corruption within governing bodies, bad governance can be reduced. There are three ways to reduce corruption:

Introduction of New Institutions and Laws

Corruption occurs usually due to the weak or a lack of anti-corruption laws and institutions in place. Bad governance by governing entities relies on pre-existing laws or norms which lack anti-corruption methods to get away or explain corruption to foreign criticism. In these cases, by introducing either stronger or new anti-corruption laws or institutions that change or monitor these old laws, corruption can be monitored and curbed. Starting these new institutions and laws is an attempt to push the governing bodies in the correct direction by removing the worst features first and not immediately removing corruption from the body. However, governing entities, as research has shown, are most likely to attempt adopting new institutions and laws but are forced back by withholding political tools necessary.

Target Vulnerable Services

The World Bank and macro economists tend to view corruption in governing bodies as a whole and don't take into consideration the differences in corruption that the different services within the body are susceptible too. Reforms to these individual services within a body which impose anti-corruption regulations act to curb corruption.

Change Service at Grassroots Level

Nine principles can be applied to help deal with corruption by disturbing the laws and institutions that provide the opportunity for officials to act with motives of self-gain. This disruption is initiated by national policymakers using their power to change laws and institutions.

Principle 1: Repeal Laws that Facilitate the Collection of Bribes

The laws and regulations affect the process of services provided by governing bodies. The larger the amount of stages in the process, the more opportunities governing officials have to delay and impede the process. If the entity requesting the service requires a quick claim, the most effective way to remove the delays by the governing bodies is by paying a bribe. By minimising the amount of regulations and the amount of steps in the process, governing bodies have less opportunities to impede the process.

Principle 2: Increase the Use of Objective Criteria for Deciding Eligibility for Recovering a Service

A list of criteria is needed for each service provided to determine whether or not a person is allowed to use it. To ensure that this process of determining if someone is able to use the service is impartial, applying criteria which are objective is the best method. By increasing the number of objective criteria in services that rely on evaluation, such as the education system, the room for corruption and bribery is decreased.

Principle 3: Use Computers to Reduce Contact with Public Officials

Bribery is one of the most common forms of corruption. The normal method of bribery occurs in hand-to-hand trades of cash. Such a trade means a face-to-face transaction needs to occur between the two parties. By increasing the use of computers in services, the face-to-face element is removed from the transaction. This eliminates the opportunity for public officials to benefit through bribery as they aren't presented the chance.

Principle 4: Monitor the Delivery of Services Electronically

Corruption and fraud often occur hand in hand. Public officials who collect bribes are likely to defraud their agency by being away from the office. Collecting time sheets electronically enables agencies to check computerised data to verify the work being completed by public officials.

Principle 5: Give Citizens Access to Public Records About Themselves

Public officials keep computer files relating to requests received from citizens and the process on that request. By enabling the public to have access to these files, citizens that have made requests for a service are able to monitor what is happening in the process. This will provide transparency to the process enabling citizens to take specific actions depending on the process of their request.

Principle 6: Match the Supply of Services to Entitlements and Obligations

Laws and regulations establish who is entitled to public services, such as education and healthcare. The resources to meet these entitlements are provided by the government to the public, such as teachers and nurses. The allocation of entitlements to the public by the government becomes an issue when the demand for the entitlements becomes greater than the supply. The lack of supply can mean that to meet the necessary demand the quality or standard of the services provided drop (doctors reduce the time spent on examining patients). By correctly matching the entitlements and the obligations, a drop in a quality can be avoided.

Principle 7: Expand Choice

The government holds a monopoly over some of the services it provides, such as official documentation. Other public services that they offer, such as healthcare, education and pensions, can be offered by "non-for-profit organisations and by private-sector enterprises". By expanding the choices not only can the public select an option better suited to themselves, if there is corruption within one of the sectors, alternative options are available so that the corruption can be avoided.

Principle 8: Legalise Some Public Service Payments

Legalising payments for public services that are currently free will mean that the illegal charge, the bribe, will be legalised in a similar fashion to the principle of discount airlines. The higher the price paid for the public service, the better treatment the citizen will receive for that service.

Principle 9: Align Public Laws and Informal Standards

Informal standards of public opinion do not always align with the public laws that determine how governing bodies should deliver public services. Corporations can abide by the law and use tax loopholes to avoid paying large amounts of tax and the activity isn't considered corrupt in a legal sense. In the public opinion, the same activity can be considered corrupt as corporations avoid tax for private gain.

2.2.1.3 Conceptualising Governance Strategy

Unlike the binary of good governance vs bad governance, **strategic governance** involves the interplay between governance and strategic decision-making. This can also be called "**governance strategy**".

Often, strategic governance takes the form of collaborative efforts in which private actors opt into additional self-regulatory measures while receiving support and guidance from public bodies such as international organisations or national regulatory agencies. The area of strategy, risk and governance includes the role of corporate governance within an organisation, as well as the formulation of strategies, the translation of those strategies into specific business objectives and actions and their implementation.

For corporations, strategic governance plays into the different levels of operational decision-making, with regard to budgetary and financial restraints, as well as centralised and decentralised control divisions.

2.3 Social Responsibility in Governance

There are three basic principles which, together, comprise social responsibility in activities: sustainability, accountability and transparency (Fig. 2.1). Regardless of whether we are talking about a government, governmental organisation, corporation

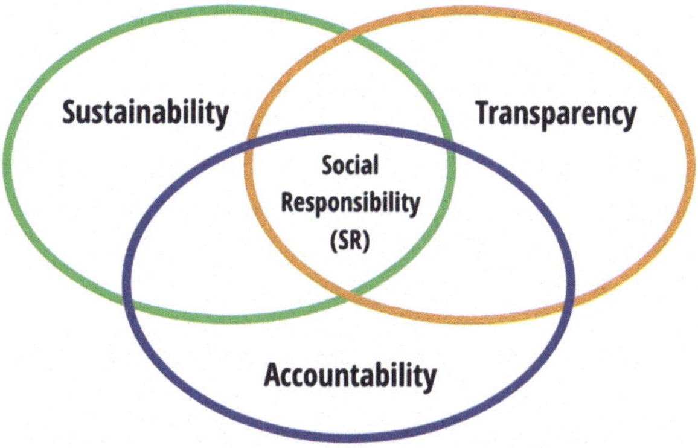

Fig. 2.1 Three facets of social responsibility

or otherwise, these three underlying principles will still very often apply. The figure below illustrates the relationship between these principles.

In this way, the three facets, sustainability, accountability, and transparency, when combined, result in great social responsibility. This can apply to governments, corporations and organisations.

2.3.1 Sustainability

Sustainability is concerned with the effect which an action taken in the present has upon the options available in the future. Consider this in terms of resources. If there are a limited amount of resources available, and those resources regenerate at a certain rate, then to use up more of the resources than are regenerated over that cycle would not be sustainable. This is all the more important when those resources are finite and non-regenerative. In our society, the principle of sustainability implies that a society must use no more of a resource than can be regenerated. This type of relationship has also been referred to as the carrying capacity of an ecosystem.[1]

We can measure the sustainability of an endeavour by considering the rate at which resources are consumed by an organisation in relation to the rate at which resources can be regenerated. The idea of a governmental organisation or corporation taking on a role as a participant in a wider social and economic system implies that these effects must be taken into account. This is necessary both for the measurement of costs and value created in the present but also taking into account the future availability of, and the demand for, those resources.

Sustainability is not just for corporations and governmental organisations but also for the governments and the citizens who live and work in that jurisdiction. Governments themselves play a key role in achieving the development goals and targets

[1] Hawken, Paul. *The Ecology of Commerce: A Declaration of Sustainability*. New York: HarperCollins Publisher, 1993.

through actions such as setting and implementing water quality policy frameworks and standards, regulating the discharge of pollutants into the environment and by taking on eco-friendly wastewater management, recycling and reuse processes. Citizens also play a role in sustainability in that the individual consumer is choosing which products or services to purchase. If these decisions are made through a sustainability lens, then the consumer has the ability to influence corporate policies in that way.

2.3.1.1 Principles of Sustainable Development

On 21 October 2015, the UN General Assembly published their *2030 Agenda for Sustainable Development*, which was adopted by all United Nations Member States in the same year. The *2030 Agenda* provides a "shared blueprint for peace and prosperity for people and the planet, now and into the future".[2]

The central features of the *Agenda* are its *17 Sustainable Development Goals* (SDGs) along with 169 specific targets, the number of which certainly feels indicative of a need for urgent action to be undertaken, by all countries, in a global partnership for sustainability.

From the Preamble to the *Agenda*:

> This Agenda is a plan of action for people, planet and prosperity. It also seeks to strengthen universal peace in larger freedom. We recognize that eradicating poverty in all its forms and dimensions, including extreme poverty, is the greatest global challenge and an indispensable requirement for sustainable development.
>
> All countries and all stakeholders, acting in collaborative partnership, will implement this plan. We are resolved to free the human race from the tyranny of poverty and want and to heal and secure our planet. We are determined to take the bold and transformative steps which are urgently needed to shift the world on to a sustainable and resilient path. As we embark on this collective journey, we pledge that no one will be left behind.
>
> The 17 Sustainable Development Goals and 169 targets which we are announcing today demonstrate the scale and ambition of this new universal Agenda. They seek to build on the Millennium Development Goals and complete what they did not achieve. They seek to realise the human rights of all and to achieve gender equality and the empowerment of all women and girls. They are integrated and indivisible and balance the three dimensions of sustainable development: the economic, social and environmental.
>
> The Goals and targets will stimulate action over the next 15 years in areas of critical importance for humanity and the planet.[3]

The *Agenda* also provides five fundamental principles on which the mandate is based. Those principles are provided in the Table 2.1. The five fundamental principles include consideration of people, planet, prosperity, peace and general.

[2]UN General Assembly, *Transforming our world: the 2030 Agenda for Sustainable Development*, 21 October 2015, A/RES/70/1, available at: https://www.refworld.org/docid/57b6e3e44.html

[3]UN General Assembly, *Transforming our world: the 2030 Agenda for Sustainable Development*, 21 October 2015, A/RES/70/1, available at: https://www.refworld.org/docid/57b6e3e44.html

Table 2.1 Fundamental principles of the 2030 Agenda

Principle	Mandate
People	We are determined to end poverty and hunger, in all their forms and dimensions, and to ensure that all human beings can fulfil their potential in dignity and equality and in a healthy environment
Planet	We are determined to protect the planet from degradation, including through sustainable consumption and production, sustainably managing its natural resources and taking urgent action on climate change, so that it can support the needs of the present and future generations
Prosperity	We are determined to ensure that all human beings can enjoy prosperous and fulfilling lives and that economic, social and technological progress occurs in harmony with nature
Peace	We are determined to foster peaceful, just and inclusive societies which are free from fear and violence. There can be no sustainable development without peace and no peace without sustainable development
Partnership	We are determined to mobilise the means required to implement this Agenda through a revitalised Global Partnership for Sustainable Development, based on a spirit of strengthened global solidarity, focused in particular on the needs of the poorest and most vulnerable and with the participation of all countries, all stakeholders and all people
General	The interlinkages and integrated nature of the Sustainable Development Goals are of crucial importance in ensuring that the purpose of the new Agenda is realised. If we realise our ambitions across the full extent of the Agenda, the lives of all will be profoundly improved and our world will be transformed for the better

The *17 Sustainable Development Goals* (SDGs) established by the UN General Assembly are listed in Table 2.2.

2.3.2 *Accountability*

Accountability is concerned with an organisation recognising that its actions affect the external environment and assuming responsibility for the effects of its actions. The implication here is a quantification of the effects and impacts of actions taken, both internal to the organisation and external. It also implies a reporting of those quantifications to all parties affected by those actions, as well as reporting to external stakeholders on the effects of actions taken by the organisation and how those actions are affecting stakeholders.

As in our discussion of sustainability, accountability also requires a recognition that the organisation or entity is part of a wider societal network and, as a result, has responsibilities to that network rather than simply to the owners or controllers of that organisation. This stance recognises that external stakeholders have the power to affect the way in which those actions of the organisation are taken. It also accepts that stakeholders take a role in whether or not such actions can be justified and if so, at what cost to the organisation and other affected stakeholders. To ensure accessibility for these reports, they must be based upon the following characteristics:

Table 2.2 The United Nations' 17 Sustainable Development Goals[a]

Goal 1	End poverty in all its forms everywhere
Goal 2	End hunger, achieve food security and improved nutrition and promote sustainable agriculture
Goal 3	Ensure healthy lives and promote well-being for all at all ages
Goal 4	Ensure inclusive and equitable quality education and promote lifelong learning opportunities for all
Goal 5	Achieve gender equality and empower all women and girls
Goal 6	Ensure availability and sustainable management of water and sanitation for all
Goal 7	Ensure access to affordable, reliable, sustainable and modern energy for all
Goal 8	Promote sustained, inclusive and sustainable economic growth, full and productive employment and decent work for all
Goal 9	Build resilient infrastructure, promote inclusive and sustainable industrialisation and foster innovation
Goal 10	Reduce inequality within and among countries
Goal 11	Make cities and human settlements inclusive, safe, resilient and sustainable
Goal 12	Ensure sustainable consumption and production patterns
Goal 13	Take urgent action to combat climate change and its impacts*
Goal 14	Conserve and sustainably use the oceans, seas and marine resources for sustainable development
Goal 15	Protect, restore and promote sustainable use of terrestrial ecosystems, sustainably manage forests, combat desertification and halt and reverse land degradation and halt biodiversity loss
Goal 16	Promote peaceful and inclusive societies for sustainable development, provide access to justice for all and build effective, accountable and inclusive institutions at all levels
Goal 17	Strengthen the means of implementation and revitalise the Global Partnership for Sustainable Development

[a]UN General Assembly, *Transforming our world: the 2030 Agenda for Sustainable Development*, 21 October 2015, A/RES/70/1, available at: https://www.refworld.org/docid/57b6e3e44.html

- Understandability—to all concerned parties
- Relevance—to the users of the information provided in the report
- Reliability—in terms of accuracy of measurement, representation of impact and freedom from bias
- Comparability—or consistency, both over time and between different organisations or entities

2.3.3 *Transparency*

Transparency means that the external impact of the actions of an organisation can be ascertained from that organisation's reporting. It also means that important facts are not disguised within that report. In effect, all of the actions of the organisation, including external impacts, should be apparent to all who use the information in the reports.

In this way transparency follows from the two previous principles in that it can be equally seen as a part of the process of recognition of responsibility—on the part of the organisation—for the external effects of its actions and equally part of the process of transferring power to external stakeholders.

2.3.4 *The Social Contract*

The **social contract theory** is a descriptive theory about society and the relationship between rules and laws and why society needs them. Over the centuries, philosophers have tried to describe the ideal social contract and to explain how existing social contracts have evolved. As a part of moral and political philosophy, the social contract theory gained prominence during the Age of Enlightenment—the mid-seventeenth to the early nineteenth centuries—when it emerged as the leading doctrine of political legitimacy.

2.3.4.1 The Origins of the Social Contract

One of the earliest descriptions of the social contract was posed by Glaucon, as described by Plato in *The Republic, Book II.* This concept also appears in Plato's dialogue, *Crito*. From *The Republic*:

> They say that to do injustice is, by nature, good; to suffer injustice, evil; but that the evil is greater than the good. And so when men have both done and suffered injustice and have had experience of both, not being able to avoid the one and obtain the other, they think that they had better agree among themselves to have neither; hence there arise laws and mutual covenants; and that which is ordained by law is termed by them lawful and just. This they affirm to be the origin and nature of justice;—it is a mean or compromise, between the best of all, which is to do injustice and not be punished, and the worst of all, which is to suffer injustice without the power of retaliation; and justice, being at a middle point between the two, is tolerated not as a good, but as the lesser evil, and honoured by reason of the inability of men to do injustice. For no man who is worthy to be called a man would ever submit to such an agreement if he were able to resist; he would be mad if he did. Such is the received account, Socrates, of the nature and origin of justice.[4]

While Plato is often seen as the earliest proponent of the social contract, other formulations of the social contract have been preserved in many of the world's oldest records, including the Indian Buddhist text of the second century BCE, *Mahāvastu*, recounts the legend of Mahasammata. The story goes as follows:

---

[4] *The Republic, Book II.* Quoted from http://classics.mit.edu/Plato/republic.3.ii.html

> In the early days of the cosmic cycle mankind lived on an immaterial plane, dancing on air in a sort of fairyland, where there was no need of food or clothing, and no private property, family, government or laws. Then gradually the process of cosmic decay began its work, and mankind became earthbound, and felt the need of food and shelter. As men lost their primaeval glory, distinctions of class arose, and they entered into agreements with one another, accepting the institution of private property and the family. With this theft, murder, adultery, and other crime began, and so the people met together and decided to appoint one man from among them to maintain order in return for a share of the produce of their fields and herds. He was called 'the Great Chosen One' (Mahasammata), and he received the title of raja because he pleased the people.[5]

The social contract theory became more widespread after Epicurus (341–270 BC)—the first philosopher who saw justice as a social contract, and not as existing in nature due to divine intervention—decided to bring the theory to the forefront of his society.[6] Epicurus combined the social contract with his conceptions of justice and law being rooted in mutual agreement and advantage.[7] This is evidenced by these lines, among others, from his *Principal Doctrines*[8]*:*

> 31. Natural justice is a pledge of reciprocal benefit, to prevent one man from harming or being harmed by another.
> 32. Those animals which are incapable of making binding agreements with one another not to inflict nor suffer harm are without either justice or injustice; and likewise for those peoples who either could not or would not form binding agreements not to inflict nor suffer harm.
> 33. There never was such a thing as absolute justice, but only agreements made in mutual dealings among men in whatever places at various times providing against the infliction or suffering of harm.[9]

2.3.4.2 The Hobbesian Social Contract

The first modern philosopher to articulate a detailed contract theory was Thomas Hobbes (1588–1679).[10] Hobbes theorised that, in the absence of political order and law, everyone would have unlimited natural freedoms, including the "right to all things" and thus the freedom to plunder rape and murder; there would be an endless "war of all against all". According to Hobbes, the lives of individuals in the state of

---

[5]AL Basham, *The Wonder That Was India*, p. 83.

[6]Huby, Pamela M. (1978). "Epicurus' Attitude to Democritus". *Phronesis*. 23 (1): 80–86. JSTOR 4182030. https://www.jstor.org/stable/4182030

[7]Long, A. A. (1999), "Epicurus", *A Companion to Philosophers*, Malden, Massachusetts and Oxford, England: Wiley Publishers, Ltd., pp. 239–243. Available at: https://books.google.com/books?id=PY5ExJZRq8kC&q=A+Companion+to+Philosophers

[8]Epicurus, and Eugene Michael O'Connor. 1993. *The essential Epicurus: letters, principal doctrines, Vatican sayings, and fragments*. Buffalo, N.Y.: Prometheus Books.

[9]Vincent Cook (2000-08-26). "Principal Doctrines". Epicurus. Available at: http://www.epicurus.net/en/principal.html

[10]Williams, Garrath. "Thomas Hobbes: Moral and Political Philosophy". *Internet Encyclopedia of Philosophy*. https://www.iep.utm.edu/hobmoral/

nature were "solitary, poor, nasty, brutish and short", a state in which self-interest and the absence of rights and contracts prevented the "social", or society. Life was "anarchic" (without leadership or the concept of sovereignty). Individuals in the state of nature were apolitical and asocial. This state of nature is followed by the social contract.[11]

Hobbes posited that to avoid an apolitical and asocial state of nature, free men contract with each other to establish a political community or civil society through a social contract in which they all gain security in return for subjecting themselves to an absolute sovereign, one man or an assembly of men. Hobbes asserted that human's consent to abdicate their rights in favour of the absolute authority of government, whether monarchical or parliamentary. The social contract was seen as an "occurrence" during which individuals came together and ceded some of their individual rights so that others would cede theirs. This results in the establishment of the state, a sovereign entity like the individuals now under its rule used to be, which would create laws to regulate social interactions. In this way, human life was thus no longer "a war of all against all".

Though the edicts of the sovereign may be arbitrary and tyrannical, Hobbes saw absolute government as the only alternative to the terrifying anarchy of a state of nature.[12] In *Leviathan*, Hobbes wrote that we, as humans, need the "terror of some power" in order to heed the law of reciprocity, that is, "doing to others, as we would be done to".[13]

2.3.4.3 The Social Contract According to Locke and Rousseau

Locke and Rousseau argued that we gain civil rights in return for accepting the obligation to respect and defend the rights of others, giving up some freedoms to do so.

John Locke's conception of the social contract differed from Hobbes' in several fundamental ways, retaining only the central notion that persons in a state of nature would willingly come together to form a state. Locke believed that individuals in a state of nature would be bound morally, by the law of nature, not to harm each other in their lives or possessions. Without a government to defend them against those seeking to injure or enslave them, Locke further believed people would have no security in their rights and would live in fear. Individuals, to Locke, would only agree to form a state that would provide, in part, a "neutral judge", acting to protect the lives, liberty and property of those who lived within it.[14]

[11] Robertson, George Croom; Anonymous texts (1911). "Hobbes, Thomas". In Chisholm, Hugh (ed.). *Encyclopædia Britannica*. Vol. 13 (11th ed.). Cambridge University Press. pp. 545–552.

[12] Gaskin. "Introduction". *Human Nature and De Corpore Politico*. Oxford University Press.

[13] Hobbes, Thomas (1985). *Leviathan*. London: Penguin. p. 223.

[14] Gaba, Jeffery (Spring 2007). "John Locke and the Meaning of the Takings Clause". *Missouri Law Review*. **72** (2).

While Hobbes argued for near-absolute authority, Locke argued for inviolate freedom under law in his *Second Treatise of Government.*[15] Locke argued that a government's legitimacy comes from the citizens' delegation to the government of their absolute right of violence, along with elements of other rights, as necessary to achieve the goal of security through granting the state a monopoly of violence, whereby the government, as an impartial judge, may use the collective force of the populace to administer and enforce the law, rather than each man acting as his own judge, jury and executioner—the condition in the state of nature.[16]

While Locke believed that natural rights were inalienable, and therefore the rule of God superseded government authority, Rousseau believed that democracy (majority rule) was the best way to ensure welfare while maintaining individual freedom under the rule of law. Jean-Jacques Rousseau (1712–1778), in his influential 1762 treatise *The Social Contract*, outlined a different version of social contract theory, as the foundations of society based on the sovereignty of the "general will".[17]

Rousseau's political theory differs in important ways from that of Locke and Hobbes.[18] Rousseau's collectivist conception is most evident in his development of the "luminous conception" of the "general will", that is, the power of all the citizens' collective interest, rather than their individual interests. Rousseau believed that society was only legitimate when the sovereign were the sole legislators. He stated that the individual must accept "the total alienation to the whole community of each associate with all his rights" and that in order for the social contract to work, individuals *must* forfeit their rights to the whole so that such conditions were "equal for all".[19]

Specifically:

> [The social contract] can be reduced to the following terms: Each of us puts his person and all his power in common under the supreme direction of the general will; and in a body, we receive each member as an indivisible part of the whole.[20]

While the Lockean concept of the social contract was invoked in the US Declaration of Independence, social contract theories, in general, were eclipsed in the

[15] Locke, J. (1980). Second treatise of government (C. B. Macpherson, Ed.). Hackett Publishing.

[16] Tuckness, Alex, "Locke's Political Philosophy", *The Stanford Encyclopedia of Philosophy* (Winter 2020 Edition), Edward N. Zalta (ed.). Available at: https://plato.stanford.edu/archives/win2020/entries/locke-political/

[17] Wokler, Robert (2001). *Rousseau: A Very Short Introduction*. Oxford: Oxford University Press. ISBN 978–0–19-280,198-2. Archived from the original on 14 August 2021. Available at: https://books.google.com/books?id=6XcRDAAAQBAJ&q=rousseau+very+short+introduction

[18] Entry, "Rousseau" in the *Routledge Encyclopedia of Philosophy*, Edward Craig, editor, Volume Eight, p. 371.

[19] Rousseau, Jean-Jacques (2002). *The social contract; and, the first and second discourses/Jean-Jacques Rousseau; edited and with an introduction by Susan Dunn; with essays by Gita May [and others]*. New Haven: Yale University Press. p. 163.

[20] Jean-Jacques Rousseau, *Oeuvres complètes*, ed. B. Gagnebin and M. Raymond (Paris, 1959–95), III, 361; *The Collected Writings of Rousseau*, ed. C. Kelley and R. Masters (Hanover, 1990–), IV, 139.

nineteenth century in favour of utilitarianism, Hegelianism and Marxism; they were revived in the twentieth century, notably in the form of a thought experiment by John Rawls.[21]

2.3.4.4 The Raelsian Social Contract

Building on the work of Immanuel Kant with its presumption of limits on the state, John Rawls (1921–2002), in *A Theory of Justice* (1971), proposed a contractarian approach whereby rational people in a hypothetical "original position" would set aside their individual preferences and capacities under a "veil of ignorance" and agree to certain general principles of justice and legal organisation. This idea is also used as a game-theoretical formalisation of the notion of fairness.[22]

The resultant theory was challenged and refined several times in the decades following its original publication in 1971. A significant reappraisal was published in the 1985 essay "Justice as Fairness" and the 2001 book *Justice as Fairness: A Restatement* in which Rawls further developed his two central principles for his discussion of justice. Together, they dictate that society should be structured so that the greatest possible amount of liberty is given to its members, limited only by the notion that the liberty of any one member shall not infringe upon that of any other member. Secondly, inequalities—either social or economic—are only to be allowed if the worst off will be better off than they might be under an equal distribution. Finally, if there is such a beneficial inequality, this inequality should not make it harder for those without resources to occupy positions of power—for instance, public office.[23]

In his work, Rawls develops what he claims are principles of justice through the use of an artificial device he calls the Original Position, in which, everyone decides principles of justice from behind a "veil of ignorance". This "veil" is one that essentially blinds people to all facts about themselves so they cannot tailor principles to their own advantage:

> [N]o one knows his place in society, his class position or social status, nor does anyone know his fortune in the distribution of natural assets and abilities, his intelligence, strength, and the like. I shall even assume that the parties do not know their conceptions of the good or their special psychological propensities. The principles of justice are chosen behind a veil of ignorance.[24]

---

[21] The Veil of Ignorance (sometimes referred to as "the original position") is a thought experiment popularised by a twentieth-century philosopher John Rawls with the goal of thinking more clearly and impartially about the fair organising principles of a society based on solidarity.

[22] Follesdal Mertens, Andreas (2005). *Real World Justice: Grounds, Principles, Human Rights, and Social Institutions*. Dordrecht: Springer. p. 88.

[23] Voice, Paul (2011). *Rawls explained: from fairness to utopia*. Open Court. pp. 41–48.

[24] Rawls, John (1999). *A Theory of Justice: Revised Edition*.

According to Rawls, ignorance of these details about oneself will lead to principles that are fair to all. If an individual does not know how he will end up in his own conceived society, he is likely not going to privilege any one class of people, but rather develop a scheme of justice that treats all fairly. In particular, Rawls claims that those in the Original Position would all adopt a maximin strategy which would maximise the prospects of the least well-off.

> They are the principles that rational and free persons concerned to further their own interests would accept in an initial position of equality as defining the fundamentals of the terms of their association.[25]

Rawls bases his Original Position on a "thin theory of the good" which he says "explains the rationality underlying choice of principles in the Original Position". A full theory of the good follows after we derive principles from the Original Position. Rawls claims that the parties in the Original Position would adopt two such principles, which would then govern the assignment of rights and duties and regulate the distribution of social and economic advantages across society. In the revised version of his work, Rawls clarifies his position on the two principles of justice: the greatest equal liberty principle and the difference principle:

1. Each person is to have an equal right to the most extensive total system of equal basic liberties compatible with a similar system of liberty for all.
2. Social and economic inequalities are to be arranged so that they are both:
   (a) To the greatest benefit of the least advantaged, consistent with the just savings principle, and
   (b) Attached to offices and positions open to all under conditions of fair equality of opportunity.[26]

2.3.4.5 The Social Contract Model

Put simply, for our purposes, the social contract theory is based upon the idea that people live together in society in accordance with an agreement that establishes moral and political rules of behaviour. Some people believe that if we live according to a social contract, we can live morally by our own choice and not because a divine being requires it. Social contract arguments typically are that individuals have consented, either explicitly or tacitly, to surrender some of their freedoms and submit to the authority (of the ruler or to the decision of a majority) in exchange for protection of their remaining rights or maintenance of the social order. The relation between natural and legal rights is often a topic of social contract theory.

With regard to the application of the social contract to social responsibility, we consider the impact of the decisions and actions made by a government or organisation and the impacts of effects of that decision or action on other stakeholders or

[25] Rawls, John (1971). *A Theory of Justice*. p. 11.

[26] Rawls, John (1999). *A Theory of Justice: Revised Edition*. p. 266.

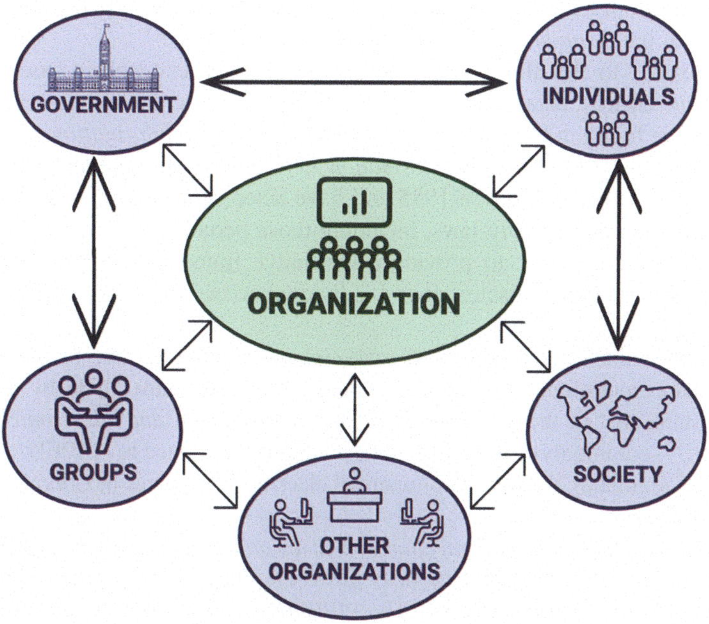

Fig. 2.2 The social contract model

related parties, that is, the impact of the organisation (or other entity) on government, individuals, groups, other organisations and society as a whole (Fig. 2.2).

2.3.5 *Comparative Provisions*

Comparatively, we can also look at the governance provisions given which aim to protect the various rights of individuals and organisations in a governing state. On a large-scale national level, we can look to specific countries, such as Canada, the United States and Australia.

2.3.5.1 Canada

In Canada, there are different statutory privacy protections which separately apply to the government, to corporations and to individuals. As well, regardless of the category into which an individual, corporation, organisation or other entity falls, all parties to whom Canadian law is applicable must abide by, and operate within, the provisions of the *Canadian Charter of Rights and Freedoms.*[27]

With respect to the ability of the government to access, collect, use, retain, disclose and destroy data within the federal public sector, the relevant statutory provisions are found in the *Privacy Act*[28] and the *Access to Information Act.*[29] The *Privacy Act*[30] and the *Access to Information Act*[31] were both implemented by the Canadian federal government in 1985 and have acted as a starting point for more recent legislation and privacy laws, including those pertaining to the cyber sector. These Acts work together to provide a legislative framework for personal data collection, use, retention, disclosure and individual access within the federal public sector.

For corporations and commercial marketplaces, including private businesses, commercial corporations, and similar organisations, they must abide by the provisions established in the *Personal Information Protection and Electronic Documents Act*[32]—commonly, and highly understandably, shortened to "PIPEDA"—and by the rules regulating the use of commercial electronic messages in *Canada's Anti-Spam Legislation.*[33]

The *Personal Information Protection and Electronic Documents Act,* or PIPEDA, officially became law in 2000 to help grow consumer trust in both electronic commerce and the digital economy. The PIPEDA provisions apply specifically to private-sector organisations, which are operating either fully or partially in Canada, and that collect, use or disclose personal information in the course of commercial activities. These provisions aim to protect the privacy of those individuals, specific subsets and targeted groups of individuals or organisations from whom the personal information has been gathered. Within the scope of this Act, a **commercial activity** is any particular transaction, act or conduct or any regular course of conduct that is of a commercial character. This also includes the selling, bartering or leasing of donors, membership or other fundraising lists.[34]

Canada's Anti-Spam Legislation, or CASL, is the federal law that addresses spam and other electronic threats. It established the rules for sending commercial electronic messages and the installation of computer programs.[35] **Spam**, in relation to CASL, refers to unwanted or unsolicited commercial electronic messages received over the Internet. As defined, a **commercial electronic message**, or CEM, is an

---

[27] *Part I of the Constitution Act, 1982. Canadian Charter of Rights and Freedoms*, s 7, Part 1 of the *Constitution Act*, 1982, being Schedule B to the Canada Act 1982 (UK), 1982, c 11.

[28] *Privacy Act* (RSC 1985, c P-21).

[29] *Access to Information Act* (RSC 1985, c A-1).

[30] *Privacy Act* (RSC 1985, c P-21).

[31] *Access to Information Act* (RSC 1985, c A-1).

[32] *Personal Information Protection and Electronic Documents Act* (SC 2000, c 5).

[33] *Canada's Anti-Spam Legislation* (SC 2010, c 23).

[34] *Personal Information Protection and Electronic Documents Act* (SC 2000, c 5).

[35] *Canada's Anti-Spam Legislation* (SC 2010, c 23).

electronically distributed and received message that encourages consumers to participate in a commercial activity. This could come in the form of an automated email containing a coupon or discount code, or it could perhaps be an SMS notifying subscribing customers about a promotion or sale.

Canada's Anti-Spam Legislation aims to protect consumers and businesses from the misuse of digital technology, including spamming and other non-consensual activities.[36] It applies to all electronic messages sent by businesses and organisations in connection with a commercial activity—that is, electronic messages sent in hopes of encouraging engagement from the consumer, with the ultimate purpose being to make a profit. The key distinguishing feature of this legislation is the requirement that Canadian—and global—organisations that send commercial electronic messages within, from or to Canada must receive express consent from recipients prior to sending those messages.[37]

Over the past couple of years, there has been some seemingly significant governmental interest shown in creating newer, more broadly applicable, digital online content regulations and mandatory content exposure standards that would specifically apply to Canadian broadcasters operating online.[38] Existing Canadian laws apply both to Canadian companies and to international companies which conduct their business endeavours within Canada. These types of platforms—those which conduct commercial endeavours in Canada—are already subject to well-established Canadian law, including the organisational provisions established under the PIPEDA and *Canada's Anti-Spam Legislation,*[39] among other examples.

There has been some recent political support in Canada for forcing platforms like Facebook, Google, Amazon and Netflix to be subject to Canadian laws and pay Canadian taxes.[40] Support has been shown for a variety of provisions, including those which would require companies which operate in Canada, including via the Internet, to obey the (currently unwritten) laws of Canada's new—but not yet existent—*Digital Charter.*

2.3.5.2 United States

In the United States, federal government and governmental agencies are bound by the *Privacy Act of 1974*[41] and the *Federal Information Security Management Act*

[36] *Canada's Anti-Spam Legislation* (SC 2010, c 23).

[37] *Canada's Anti-Spam Legislation* (SC 2010, c 23).

[38] This would not necessarily be limited to digital content streaming services but could potentially be extended to include some social networking platforms under the "broadcaster" umbrella.

[39] *Canada's Anti-Spam Legislation* (SC 2010, c 23).

[40] See the attempt by the federal liberal government to create new Internet censorship (and more!) regulations in their proposed Bill C-10, which was stalled in the senate prior to the 2021 federal election and which is now undergoing a highly secretive consultation and review process about which the public is (surprisingly?) being given extremely little information.

[41] *Privacy Act of 1974*, 5 USC § 552a.

(FISMA).[42] Additionally, the United States has sector-specific law which applies to healthcare as well as banking and finance.

The *Health Insurance Portability and Accountability Act of 1996* (HIPAA) is a US federal law that required the creation of national standards to protect sensitive patient health information from being disclosed without the patient's consent or knowledge.[43] This was amended to include the *Privacy Rule* and the *Security Rule* to enable implementation of the requirements of the HIPAA.[44]

Similarly, in finance, the *Gramm–Leach–Bliley Act* (GLBA), also known as the *Financial Services Modernization Act of 1999*, requires a wide range of financial institutions to adequately explain their information-sharing practices to their customers and to safeguard sensitive personal data. The three major components overlaying the collection, disclosure, and protection of consumers' personal information in the *Gramm–Leach–Bliley Act* include the *Financial Privacy Rule*, the *Safeguards Rule* and the *Protection from Pretexting*.[45]

2.3.5.3 Australia

Within the Australian Commonwealth, cybersecurity laws for government and corporations are guided by the *Privacy Act 1988*, the *Privacy Amendment 2012* and the *Privacy Regulation 2013*.[46] The 13 *Australian Privacy Principles* included in the *Privacy Act* legislation are to be applied in guiding the development of privacy protocols in governmental bodies, corporations and organisations which are subject to the *Privacy Act*. The use of commercial electronic messages is regulated by the *Spam Act 2003*.[47]

As the foundational piece of Australian legislation that protects the handling of personal information, Australia's *Privacy Act 1988* provides the national statutory law pertaining to the collection, use, storage and disclosure of personal information in the federal public sector as well as in the private sector. The *Privacy Act* and the 13 *Australian Privacy Principles* included within it apply to all organisations which carry out business in Australia and which include actively collecting personal information.[48]

The *Australian Privacy Principles* deal with all stages of the processing of personal information. They set out the standards for the collection, use, disclosure, quality and security of personal information, and they provide obligations

[42] *Federal Information Security Management Act of 2002*, 44 USC § 3541.

[43] *Health Insurance Portability and Accountability Act*. Pub. L. No. 104–191, § 264, 110 Stat.1936.

[44] *Health Insurance Portability and Accountability Act*. Pub. L. No. 104–191, § 264, 110 Stat.1936.

[45] *Gramm–Leach–Bliley Act* (GLBA), 15 USC. § 6801 et seq.

[46] *Privacy Act 1988 (Cth).*

[47] *Spam Act 2003 (Cth).*

[48] *Privacy Act 1988 (Cth).*

concerning access to, and correction of, an individual's own personal information for agencies and organisations which are subject to the *Privacy Act*.

These 13 principles are intended to protect individual privacy while simultaneously not overburdening agencies and organisations with inflexible, and potentially expensive, prescriptivist rules. The *Australian Privacy Principles*, and the bodies which fall under, are overseen by the Office of the Australian Information Commissioner within the Australian Government.[49]

The *Privacy Act* is further supported by the *Privacy Amendment (Enhancing Privacy Protection) Act 2012*[50] and the *Privacy Regulation 2013*. In February 2018, the *Notifiable Data Breaches* scheme was implemented as part of the *Privacy Act*, requiring notification to all affected individuals and to the Office of the Australian Information Commissioner when any party, who is subject to the *Privacy Act*,[51] experiences a data breach of personal information which is likely to pose a risk of serious harm to the affected individuals.

The different relevant laws which apply to corporations and specific sectors in Canada, the United States and Australia are summarised in the Table 2.3.

2.4 Digital Governance

In relation to social responsibility, **digital governance** requires the acceptance by the government, corporation or organisation of the effects of actions undertaken directly or indirectly online. Digital governance is a framework for establishing accountability, roles, decision-making and change management authority for an organisation's digital presence. Having a well-designed digital governance framework minimises effort and cost and ensures digital business maturity.

2.4.1 Elements

With regard to digital governance, there are many stakeholders. However, we can narrow these stakeholders down to three main parties: government, citizens and businesses (Fig. 2.3).

---

[49] The Office of the Australian Information Commissioner is the independent national regulator for privacy and freedom of information in Australia. Its role is to promote and uphold the right of individuals and entities to access government-held information and the right to have their personal information protected while it is collected, used, held or otherwise.

[50] https://www.aph.gov.au/Parliamentary_Business/Bills_Legislation/bd/bd1213a/13bd020#:~:text=The%20purpose%20of%20the%20Privacy,Privacy%20Law%20and%20Practice'%20(ALRC

[51] *Privacy Act 1988 (Cth).*

Table 2.3 Private sector laws of Canada, the United States and Australia

Country	Canada	United States	Australia
All private corporations	• *Canadian Charter of Rights and Freedoms* (1982) • *Personal Information Protection and Electronic Documents Act* (2000) (PIPEDA)	• *Privacy Act of 1974* • *other applicable laws are based on state-specific legislation	• *Privacy Act (1988)* • *Privacy Amendment (Enhancing Privacy Protection) Act (2012)* • *Australian Government Agencies Privacy Code (2017)*
Healthcare specific	• *Canadian Charter of Rights and Freedoms* (1982) • *Personal Information Protection and Electronic Documents Act* (2000) (PIPEDA)	• *Health Insurance Portability and Accountability Act of 1996* (HIPAA)	• *Privacy Act (1988)* • *Privacy Amendment (Enhancing Privacy Protection) Act (2012)*
Banking and finance specific	• *Canadian Charter of Rights and Freedoms* (1982) • *Personal Information Protection and Electronic Documents Act* (2000) (PIPEDA)	• *Gramm–Leach–Bliley Act* (GLBA) aka *Financial Services Modernization Act of 1999*	• *Privacy Act (1988)* • *Privacy Amendment (Enhancing Privacy Protection) Act (2012)*
Commercial electronic messages	• *Canada's Anti-Spam Legislation* (2014) (CASL)	• *Controlling the Assault of Non-Solicited Pornography And Marketing Act (2003)*	• *Spam Act (2003)*
Copyright	• *Copyright Act* (1985)	• *Copyright Act of 1976*	• *Copyright Act (1968)*

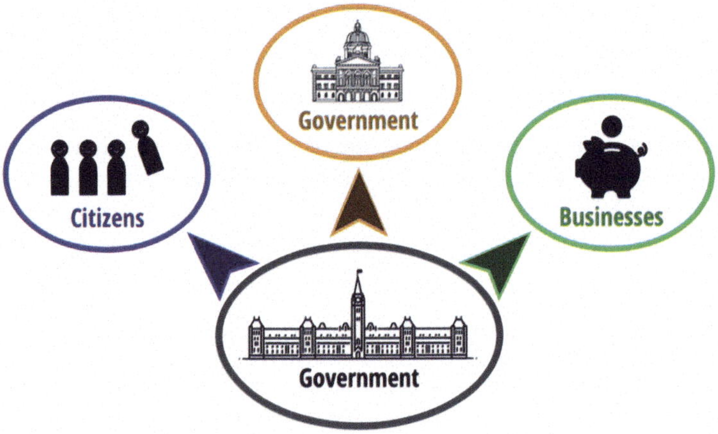

Fig. 2.3 Governance relationships—G2C, G2B, G2G

2.4.2 *Types of Electronic Governance*

Electronic governance or **e-governance** is the application of information technology for delivering government services, exchange of information, communication transactions, integration of various stand-alone systems between government-to-citizen (G2C), government-to-business (G2B), government-to-government (G2G) and government-to-employees (G2E) as well as back-office processes and interactions within the entire governance framework. Through e-governance, government services are made available to citizens through IT.[52]

According to United Nations' definition, e-government comprises three integral parts[53]:

- **Government-to-citizen (G2C)** programs are meant to make it easier for citizens and consumers of public services to communicate with the government. This covers interactions, including the delivery of public services and participation in consultation and decision-making processes.
- **Government-to-business (G2B)** refers to business-specific transactions (such as payments, selling and acquisition of goods and services, etc.) and the provision of business-focused services online.
- **Government-to-government (G2G)** involves sharing data and conducting electronic communications between government agencies. This consists of intra- and inter-agency interactions at the national level and exchanges between the national, provincial and local levels.

Other models contain an additional (fourth) category, that of government-to-employees:

- **Government-to-employees (G2E)** includes maintaining personal information and employee records. It also involves e-payroll (to view paychecks, pay stubs, pay bills and keep records for tax information) and e-learning (to keep employees informed on the important materials they need to know through the use of visuals, animation or videos via a computer-based learning tool).

2.4.2.1 Government-to-Business (G2B)

Government-to-business (G2B) is a relationship between businesses and government, where government agencies of various levels provide services or information to a business entity via government portals or with the help of other IT solutions.

[52] Saugata, B., and Masud, R.R. (2007). *Implementing E-Governance Using OECD Model (Modified) and Gartner Model (Modified) Upon Agriculture of Bangladesh*. IEEE. 1–4244–1551-9/07.

[53] https://publicadministration.un.org/egovkb/en-us/about/unegovdd-framework

Such assistance from the government bodies' side aims to facilitate swift and smooth business operations, as well as a fair and transparent environment to do business.

The range of the G2B services is broad and include, among many others:

- Online information and advisory services
- Government contracting
- Digital procurement marketplaces
- Business licences, permits and regulation updates
- Electronic auctions
- Tax, social insurance payments and reporting
- Electronic forms
- Online application submission functionalities
- Virtual business dispute resolution
- Online startup registering
- Data centres, SaaS, PaaS or IaaS for e-government use

In the G2B model, the initiative comes from a government agency, and businesses are the target audience. In fact, businesses play the role of customers in the e-commerce model, and government agencies are, figuratively speaking, the customer support service.

In the globalised world, governments who succeeded in digitising their services and creating the most comfortable conditions for businesses and more fair market value will receive more foreign direct investment.

The G2B model of interaction between government and business has been developing rapidly in recent years, ensuring a wide use of digital technologies across the economies and bringing benefits to everyone, from multinationals to small- and medium-sized businesses.

2.4.2.2 Business-to-Government (B2G)

Unlike the G2B model, in B2G (business-to-government) relationships, the initiative comes from business. The official definition says that B2G is "the sale and marketing of goods and services to federal, state, or local agencies," and B2G sales and marketing is oftentimes referred to as public sector marketing.[54]

Most contracts between government agencies and companies are signed in response to an agency's request for proposal (RFP). Businesses, in their turn, bid for contracts by submitting RFP responses. The bidding process can take place online in real time and be highly competitive.

Considering the wide range of federal, state and local purchasing requirements, an entire Internet sector has developed to help match businesses to government agencies.

[54] https://www.investopedia.com/terms/b/business-to-government.asp

2.5 Corporate Governance

Corporate governance is a system of rules, policies and practices that dictate how a company's board of directors manages and oversees the operations of a company.[55] Corporate governance includes principles of transparency, accountability and security.[56] Corporate governance is the collection of mechanisms, processes and relations used by various parties to control and to operate a corporation. There are seven main characteristics of corporate governance.[57] They are as follows[58]:

1. **Discipline**
   Corporate discipline is a commitment by a company's senior management to adhere to universally recognized and accepted behaviour to be correct and proper. This encompasses a company's awareness of and commitment to the underlying principles of good governance, particularly at the senior management level[59]

   All involved parties will have a commitment to adhere to procedures, processes, and authority structures established by the organisation.[60]

2. **Transparency**
   Transparency is the ease with which an outsider is able to make a meaningful analysis of a company's actions, its economic fundamentals, and the non-financial aspects pertinent to that business. This is a measure of how good management is at making necessary information available in a candid, accurate, and timely manner—the audit data and general reports and press releases. It reflects whether investors obtain a true picture of what is happening inside the company[61]

   All actions implemented, and their decision support will be available for inspection by an authorised organisation and provider parties.[62]

3. **Independence**
   Independence is the extent to which mechanisms have been put in place to minimise or avoid potential conflicts of interest that may exist, such as dominance by a strong chief executive or large shareholder. These mechanisms range from

[55] https://pubs.opengroup.org/architecture/togaf80-doc/arch/p4/gov/gov.htm

[56] http://docplayer.net/41585744-Togaf-the-open-group-architecture-framework-version-8-1-enterprise-edition.html

[57] https://www.estate-living.co.za/news/7-charactristsic-of-good-corporate-governance-applicable-to-a-homeowners-associations/

[58] Ramani Naidoo. 2002. *Corporate Governance: An Essential Guide for South African Companies.* Juta and Company Ltd., 2002.

[59] Ramani Naidoo. 2002. *Corporate Governance: An Essential Guide for South African Companies.* Juta and Company Ltd., 2002.

[60] https://pubs.opengroup.org/architecture/togaf80-doc/arch/p4/gov/gov.htm

[61] Ramani Naidoo. 2002. *Corporate Governance: An Essential Guide for South African Companies.* Juta and Company Ltd., 2002.

[62] https://pubs.opengroup.org/architecture/togaf80-doc/arch/p4/gov/gov.htm

the composition of the board to appointments to committees of the board and external parties such as the auditors. The decisions made and internal processes established should be objective and not allow for undue influences[63]

> All processes, decision-making, and mechanisms used will be established so as to minimise or avoid potential conflicts of interest.[64]

4. **Accountability**
   Individuals or groups in a company, who make decisions and take actions on specific issues, need to be accountable for their decisions and actions. Mechanisms must exist and be effective to allow for accountability. These provide investors with the means to query and assess the actions of the board and its committees.[65]

   > Identifiable groups within the organisation—e.g., governance boards who take actions or make decisions—are authorised and accountable for their actions.[66]

5. **Responsibility**
   With regard to management, responsibility pertains to behaviour that allows for corrective action and for penalising mismanagement. Responsible management would, when necessary, put in place what it would take to set the company on the right path. While the board is accountable to the company, it must act responsively to and with responsibility towards all stakeholders of the company[67]

   > Each contracted party is required to act responsibly to the organisation and its stakeholders.[68]

6. **Fairness**
   The systems that exist within the company must be balanced in taking into account all those that have an interest in the company and its future. The rights of various groups have to be acknowledged and respected. For example, minority shareowner interests must receive equal consideration to those of the dominant shareowner(s).[69]

   > All decisions taken, processes used, and their implementation will not be allowed to create an unfair advantage to any one particular party.[70]

---

[63] Ramani Naidoo. 2002. *Corporate Governance: An Essential Guide for South African Companies.* Juta and Company Ltd., 2002.

[64] https://pubs.opengroup.org/architecture/togaf80-doc/arch/p4/gov/gov.htm

[65] Ramani Naidoo. 2002. *Corporate Governance: An Essential Guide for South African Companies.* Juta and Company Ltd., 2002.

[66] https://pubs.opengroup.org/architecture/togaf80-doc/arch/p4/gov/gov.htm

[67] Ramani Naidoo. 2002. *Corporate Governance: An Essential Guide for South African Companies.* Juta and Company Ltd., 2002.

[68] https://pubs.opengroup.org/architecture/togaf80-doc/arch/p4/gov/gov.htm

[69] Ramani Naidoo. 2002. *Corporate Governance: An Essential Guide for South African Companies.* Juta and Company Ltd., 2002.

[70] http://www.swview.org/blog/seven-characteristics-corporate-governance#:~:text=%E2%80%9CAll%20decisions%20taken%2C%20processes%20used,to%20any%20one%20particular%20

7. **Social Responsibility**
   A well-managed company will be aware of and respond to social issues, placing a high priority on ethical standards. A good corporate citizen is increasingly seen as non-discriminatory, non-exploitative, and responsible with regard to environmental and human rights issues. A company is likely to experience indirect economic benefits such as improved productivity and corporate reputation by considering those factors[71]

We will be discussing social responsibility in much greater detail in the following section.

2.5.1 *Corporate Social Responsibility*

The intersection of corporate social responsibility and user-generated content is a challenge to navigate within the law because there has been a strong debate between the rights and ownership of the content produced and the responsibility of the corporations concerning that content. While a corporation cannot necessarily be held accountable for content created by its users, the growth of digital content hosting platforms has altered these considerations, as online enterprises begin to retain profit from the hosting of such content.

Human rights and related considerations are a crucial component of modern digital corporate social responsibility and must be weighted as such when creating or altering regulatory standards and/or relevant legislation. It is increasingly necessary for both corporations and governmental organisations to consider the wide-ranging impacts of corporate activities online, whether financial, social or content-related.

Figure 2.4 below provides the proposed facets of corporate social responsibility in our contemporary digital age, or the facets of "digital corporate social responsibility".

Collectively, these ten facets of digital corporate social responsibility represent the considerations that should go into the development of a corporate social responsibility (CSR) plan.

party.%E2%80%9D&text=A%20well%2Dmanaged%20company%20will,high%20priority%20on%20ethical%20standards

[71] Ramani Naidoo. 2002. *Corporate Governance: An Essential Guide for South African Companies*. Juta and Company Ltd., 2002.

Fig. 2.4 Facets of corporate social responsibility in a digital age

2.5.2 Objectives

The World Business Council for Sustainable Development defines corporate social responsibility (CSR) as "the continuing commitment by business to behave ethically and contribute to economic development while improving the quality of life of the workforce and their families as well as of the local community and society at large".[72]

However, the difficulty arises when we try to define what, exactly, is and is not socially responsible for a corporation. You will note that this also ties back into the idea of the social contract, which was discussed earlier in this chapter. For this reason there exist many different working definitions of corporate social responsibility.

The broadest definition of corporate social responsibility is concerned with what is—or what should be—the relationship between global corporations, governments of countries and individual citizens. On a more localised level, corporate social responsibility is concerned with the relationship between a corporation and the local society or community in which that corporation resides or operates in some form. A third definition includes a concern for the relationship between a corporation and its stakeholders. While the range of specific definitions is varied, some identifiable underlying principles tie them together.

[72] https://www.social-responsibility.at/definitions/world-business-council-for-sustainable-development-2000/

Fig. 2.5 Corporate social responsibility buzzwords

For our purposes, we will define the concept of corporate social responsibility—or CSR—as based on the premise that a business must not be concerned solely with its profitability and growth but also with its social and environmental impact. Corporate social responsibility aims to maximise shared value among organisations, employees, shareholders, customers, local stakeholders and community members. Accordingly, corporate social responsibility refers to a commitment made by a corporation to manage the effects of its operations responsibly and in line with public expectations: socially, environmentally and economically.

In Fig. 2.5 below, we can see the keywords which are affiliated with corporate social responsibility.

We can begin our exploration of corporate social responsibility by discussing the foundational principles which underlie the concept.

2.5.2.1 Foundational Principles

Generally, there are four facets of responsibility that are relevant for corporations: economic responsibility, legal responsibility, ethical responsibility and philanthropic responsibility. These are broken down in the Table 2.4.

Based on the order of priority, in theory, a corporation could begin by fulfilling their economic responsibilities; working to ensure their legal responsibilities; reviewing, evaluating, revising and maintaining their ethical responsibilities; and,

Table 2.4 Areas of social responsibility for corporations

Priority level	Type of responsibility	Inherent messaging
First	Economic responsibility	"Be profitable"
Second	Legal responsibility	"Obey the law"
Third	Ethical responsibility	"Be ethical"
Fourth	Philanthropic responsibility	"Be a good corporate citizen"

finally, aiming to achieve their philanthropic responsibilities, ensuring that each area has been covered and maintaining each level of responsibility as more levels are added.[73]

The Table 2.5 shows the distinctions between implicit corporate social responsibility frameworks and explicit corporate social responsibility frameworks.

2.5.2.2 Digital Corporate Social Responsibility

Digital corporate social responsibility extends the social obligations of a corporation to that which is connected to their position in the online realm. With regard to digital content, this can include responsibility related to the legality of that content and the mitigation of harm, corporate liability, negligence and copyright infringement.

As an example, consider the FinTech industry. FinTech—or financial technology—is the digital intersection of technology and finance. This can include all forms of transactional relationships involving an exchange of currency or other forms of value exchange. The implications associated with FinTech range widely, as do those which apply in the offline world.

As another example, we can look at the recent issue encountered by Dyson with regard to one of its suppliers. Following an audit of its labour practices and allegations of rights violations by a whistleblower, Dyson cut ties with ATA, their largest Malaysian supplier. As a result, ATA has agreed to work with Malaysia's National Human Rights Commission to improve its environmental, labour and social and governance practices.

Certainly, in a modern digital age, it is increasingly necessary for corporations to consider each facet of social responsibility through a digital lens while also being aware of their own digital footprint and the potential for human rights implications through that footprint. The inclusion of human rights and digital presence into digital corporate social responsibility is provided in the figure below.

[73] Brown, Jill & Clark, Cynthia & Buono, Anthony. (2018). The United Nations Global Compact: Engaging Implicit and Explicit CSR for Global Governance. Journal of Business Ethics. 147. 10.1007/s10551-016-3382-5.

Table 2.5 Implicit vs explicit CSR frameworks[a]

Framework	Explicit CSR framework	Implicit CSR framework
Function	Describes corporate activities to assume responsibility for the interests of society	Describes the role of the corporation within the wider formal and informal institutions for the interests and concerns of society
Structure	Consists of voluntary corporate policies, programs and strategies	Consists of values, norms and rules which result in requirements for corporations, which are often codified and mandatory
Motivation	Incentives and opportunities are motivated by the perceived expectations of different stakeholders of the corporation	Motivated by the societal consensus on the legitimate expectations of the roles and contributions of all major groups in society, including corporations
Obligations	Obligations of actors are addressed in individual terms	Obligations of actors are addressed in collective terms
Identity schema	Firm-level identity	Collective identity
Economic structure	Liberal market economies	Coordinated market economies
Public procurement influences	Firms with low involvement	Firms with high involvement
Associated with...	Increasing discretionary orientation	Increasing legal orientation
Conceptual framework based on encouraging...	• Individualism • Discretionary agency • Incentivising responsive actors • Liberalism • Network governance • Policies providing discretion • Isolated actors	• Collectivism • Systemic/obligatory agency • Incentivising program-driven agency • Solidarity • Partnership governance • Policies providing obligations • Interlocking/associated actors

[a]Matten, Dirk & Moon, Jeremy. (2008). "Implicit" and "Explicit" CSR: A Conceptual Framework for a Comparative Understanding of Corporate Social Responsibility. Academy of Management Review. 33. 10.5465/AMR.2008.31193458

2.5.3 User-Generated/Earned Content

User-generated content (UGC) refers to any content created by the users of a product or service. User-generated content can consist of any form of digital content: written, videoed, posted, blogged, created, generated, placed, posted, uploaded or shared to a forum or online content hosting platform, by individual users rather than by commercial brands, organisations or companies. The content created by these particular users, or content creators, can be artistic, creative, copyrighted, personal, informational, etc.

There are many benefits for corporations who encourage positive user-generated content. Some of these benefits include:

- Boosting brand authenticity, awareness and credibility
- Building customer trust and brand loyalty
- Increasing user engagement
- Increasing conversion rates
- Increasing social media following and presence
- Increasing overall purchases and thereby increasing overall company profits
- Producing organically grown creative content, the variety of which may not normally be considered by a typical commercial marketing team

User-generated content occurs when customers create and disseminate online ideas about a product or the firm that markets it. These ideas are often in the text but also come in other forms such as music, photos or videos. User-generated digital content has the following key characteristics:

1. It is generated or created by users of a product or service rather than the firm that owns and/or sells the product or service.
2. It is creative, with the individual user adding something new and unique.
3. It is posted online.
4. It is generally and/or publicly accessible.

Put very simply, online user-generated digital content is any online content that is not created or directly controlled by the company or the online content hosting platform. Examples of user-generated content in relation to corporations include:

- Social media posts that mention a brand or company by the name
- Product reviews and ratings
- Images and videos of a product being used by a customer

By engaging with user-generated content, a company or brand can open up larger and more numerous communication channels, allowing for more extensive customer interaction. Examples of consumer engagement involving user-generated content include things like answering customer questions, responding to customer complaints and providing digital social spaces to encourage customers to interact and share experiences among themselves.

The value and impact of quality user-generated content can be uncovered by consulting the relevant research, which has been extensive.[74] Some of the available data collected on user-generated content so far has generated results which indicate that:

- 92% of consumers turn to people that they know for product and service referrals.[75]

[74] Get ready for a bunch of stats, in order of positive percentage from highest to lowest because why not.

[75] Sourced from data collected by the Placeit Graphic Design Blog, Ametys and Neilson.

- 92% of consumers trust online content garnered from friends and family above all other forms of brand messaging.[76]
- 92% of marketers believe that most, or all, content created by their brand resonate as authentic with consumers.[77]
- 90% more time is spent by consumers on sites that contain a gallery for user-generated content.[78]
- 90% of boomers prefer to use Facebook for social media engagement.[79]
- 90% of millennials prefer real and organic brands over perfect-looking and pre-packaged ones.[80]
- 89% increases in online conversion rates are positively associated with users interacting with consumer reviews.[81]
- 86% of businesses now use user-generated digital content in their marketing efforts.[82]
- 85% of fans of brands on Facebook recommend those brands to others.[83]
- 85% of consumers are more likely to make a purchase when they can find online recommendations to support advice received offline.[84]
- 85% of make-up shoppers who are between the ages of 18 and 34, and are based in the United States, use Instagram for beauty-related activities.[85]
- 85% of consumers find visual user-generated content to be more influential than brand-created photos and videos.[86]
- 84% of online shoppers use social media sites.[87]
- 83% of consumers are more likely to trust recommendations from peers over messaging from the brand itself.[88]
- 81% of shoppers would pay more for a product that includes positive user-generated content.[89]

[76] Sourced from data collected by the University of Brighton.

[77] Sourced from data collected by Business Wire, Stackla and others.

[78] Sourced from data collected from SmartFocus.

[79] Sourced from data collected from SmartFocus.

[80] Sourced from data collected from FinancesOnline and Stackla.

[81] Sourced from data collected by Bazaarvoice.

[82] Sourced from data collected from SmartFocus.

[83] Sourced from data collected by Ametys and noted as being sourced from Syncapse.

[84] Sourced from data collected by MAVRCK and referenced as: http://mashable.com/2014/04/09/millennials-user-generated-media

[85] Sourced from data collected by Facebook IQ.

[86] Sourced from data collected by the Placeit Graphic Design Blog.

[87] Sourced from data collected by MAVRCK and referenced as: http://mashable.com/2014/04/09/millennials-user-generated-media

[88] Sourced from data collected by Lundmark Advertising.

[89] Sourced from data collected by the Placeit Graphic Design Blog.

- 80% of make-up shoppers who are between the ages of 18 and 34, and are based in France, use Instagram for beauty-related activities.[90]
- 80% of customers opt not to make a purchase based on a negative online consumer review.[91]
- 76% of shoppers find content that is posted by other consumers to be more honest than branded content.[92]
- 76% of make-up shoppers between the ages of 18 and 34, and are based in Australia, use Instagram for beauty-related activities.[93]
- 75% of millennials and 50% of boomers prefer photos over written content or videos.[94]
- 74% of consumers rely on social media for their purchasing decisions.[95]
- 71% of millennials engage with social media daily, while only 60% of millennials watch television daily.[96]
- 70% of people trust images taken, uploaded and shared by "people like them" over brand-created images.[97]
- 69% of people have some form of social media account.[98]
- 65% of brands believe that user-generated content is more influential than brand-produced photos and videos.[99]
- 64% of consumer reviews are posted within 24 hours of the initial consumer experience.[100]
- 61% of people say that they would be more likely to engage with an advertisement if it contained user-generated content.[101]
- 60% of brand marketing and creation costs can be saved by focusing heavily on curating user-generated content.[102]

---

[90] Sourced from data collected by Facebook IQ.

[91] Sourced from data collected by MAVRCK and referenced as: http://dealer-communications.com/news/survey-4-out-of-5-consumers-reverse-purchase-decisions-based-on-negative-online-reviews

[92] Sourced from data collected by Adweek from Olapic's "Consumer Trust Survey" which worked with census researchers to poll 1000 US residents between the ages of 16 and 65 on their preferences around brand and consumer-generated social content.

[93] Sourced from data collected by Facebook IQ.

[94] Sourced from data collected by Adweek from Olapic's "Consumer Trust Survey" which worked with census researchers to poll 1000 US residents between the ages of 16 and 65 on their preferences around brand and consumer-generated social content.

[95] Sourced from data collected by the Placeit Graphic Design Blog.

[96] Sourced from data collected by the University of Brighton.

[97] Sourced from data collected by Online Sales Guide Tips.

[98] Sourced from data collected by Lundmark Advertising.

[99] Sourced from data collected by the University of Brighton.

[100] Sourced from data collected by Bazaarvoice.

[101] Sourced from data collected by Online Sales Guide Tips.

[102] Sourced from data collected from SmartFocus.

- 60% of surveyed respondents say that they have used branded hashtags.[103]
- 59% of consumers trust online reviews just as much as personal recommendations.[104]
- 54% of millennials want friends and the brand to know that they like the product.[105]
- 53% of millennials have said that user-generated content has influenced their online purchasing decisions.[106]
- 52% of consumers say that watching product videos makes them feel more confident in their purchasing decisions.[107]
- 51% of consumers believe that less than half of brands create content that resonates as being authentic.[108]
- 50% of boomers do not use hashtags.[109]
- 50% of consumers find user-generated content to be more memorable than brand-produced content.[110]
- 47% of millennials say that they trust user-generated content, in comparison to 36% of boomers.[111]
- 33% of shoppers say that brand and/or product authenticity is a top factor for consideration in purchasing decisions.[112]
- 30% of media time spent by millennials involves interaction with content created and curated by their peers.[113]

[103] Sourced from data collected by Adweek from Olapic's "Consumer Trust Survey" which worked with census researchers to poll 1000 US residents between the ages of 16 and 65 on their preferences around brand and consumer-generated social content.

[104] Sourced from data collected by MAVRCK and referenced as: http://www.emarketer.com/Article/Moms-Place-Trust-Other-Consumers/1007509

[105] Sourced from data collected by Adweek from Olapic's "Consumer Trust Survey" which worked with census researchers to poll 1000 US residents between the ages of 16 and 65 on their preferences around brand and consumer-generated social content.

[106] Sourced from data collected by the University of Brighton.

[107] Sourced from data collected by MAVRCK and referenced as: http://blog.filemobile.com/ugc-statistics

[108] Sourced from data collected by Business Wire, Stackla and others.

[109] Sourced from data collected by Adweek from Olapic's "Consumer Trust Survey" which worked with census researchers to poll 1000 US residents between the ages of 16 and 65 on their preferences around brand and consumer-generated social content.

[110] Sourced from data collected by the University of Brighton.

[111] Sourced from data collected by Adweek from Olapic's "Consumer Trust Survey "which worked with census researchers to poll 1000 US residents between the ages of 16 and 65 on their preferences around brand and consumer-generated social content.

[112] Sourced from data collected from Finances Online and Stackla.

[113] Sourced from data collected by the University of Brighton.

- 25% of millennials say that they trust content created by brands, in comparison to 24% of boomers.[114]
- 25% of search results for the top 20 brands in the world are links to user-generated content platforms.[115]
- 25% of millennials say that they would rank Instagram as their favourite social media platform.[116]
- 25% of consumer questions are answered within 24 hours, with a median consumer question response time of 2.8 days.[117]
- 17% of consumer reviews are in languages other than English.[118]

The many benefits derived from user-generated content, for the content host or platform, are both clear and vast. These benefits include low-cost promotional material, positive impacts on product demand, increased sales and fresh media marketing content.

The benefit to the contributors, however, is much less transparent than those for the host. There are various theories behind the motivation for contributing user-generated content, which range from altruism to social motivation, to materialistic perspectives.

As a result of the extremely high value of user-generated content in marketing, many commercial websites have opted to use incentivising strategies to encourage the generation of high-quality user-generated content by consumers. Common easy examples of incentivised user-generated content include the use of branded hashtags; entry into contests; rewards, points, freebies and other incentives for providing product reviews; and active ongoing encouragement for customers to submit photos or videos of the product. Occasionally, users may also be offered monetary incentives to positively reinforce brand loyalty and to encourage consumers to create high-quality and unique user-generated digital content, which can then be used by the brand for advertising purposes.[119]

[114]Sourced from data collected by Adweek from Olapic's "Consumer Trust Survey" which worked with census researchers to poll 1000 US residents between the ages of 16 and 65 on their preferences around brand and consumer-generated social content.

[115]Sourced from data collected by MAVRCK and referenced as: http://pressroom.ups.com/pressroom/staticfiles/pdf/fact_sheets/2013_PulseShopper_FINAL.pdf. Also referenced by Ametys and sourced in Kissmetrics.

[116]Sourced from data collected by Adweek from Olapic's "Consumer Trust Survey" which worked with census researchers to poll 1000 US residents between the ages of 16 and 65 on their preferences around brand and consumer-generated social content.

[117]Sourced from data collected by Bazaarvoice.

[118]Sourced from data collected by Bazaarvoice.

[119]Toluna:"Mixing Financial, Social and Fun Incentives for Social Voting" archived from the original PDF on 9 July 2011.

Table 2.6 Explicit vs implicit consumer incentives

Explicit incentives for consumers to create user-generated content	Implicit incentives for consumers to create user-generated content
Primary characteristic: • Tangibility	Primary characteristic: • Intangibility
Externality rewards: • Tangible rewards • Financial incentives • Economic or material incentives • Entry into a contest. • Voucher or coupon. • Point-reward system • Frequent flyer miles • Rebates and cash-back opportunities	Social connection: • Relationship development between users • Positive reinforcement of community membership • Increased density of connections between users • Social sharing of personal media • Social sharing of experiences • Social incentive • Social comparison

These types of consumer incentives can generally all be categorised into one of two categories: implicit incentives and explicit incentives.[120] These two distinct categories of consumer content creation incentive—those being explicit and implicit—are further broken down in the following Table 2.6.

2.5.4 *Legal Considerations*

The legal issues which intersect with and relate to user-generated content can be wide-ranging: from determination of ownership and copyright to personal privacy protection and individual freedom of expression to issues of cyberbullying, illegal content distribution and large-scale corporate data breaches.

Consider, for example, the facts of the following hypothetical scenario: MeTube—creators, consumer and hosts.

2.5.4.1 Example: MeTube: Creators, Consumers and Hosts

Facts:

1. MeTube is an online video hosting platform which provides a forum for users to upload video content and have that content shared with other like-minded users. Correspondingly, the platform also provides access to user-uploaded content for registered users.

[120]Chen, Yan; Harper, F. Maxwell; Konstan, Joseph; Li, Sherry Xin (September 2010). "Social Comparisons and Contributions to Online Communities: A Field Experiment on MovieLens". *American Economic Review*. **100** (4): 1358–1398.

2. All visitors to the website must be a registered user before being able to upload or view content to the hosting platform.
3. Video content on MeTube can be either general or specialised. While access to the general content is free for all registered users, access to the specialised content is only available for registered users who have also subscribed to a specialised access payment plan.

   (a) To access free content, users first must register with the hosting platform, via the website, and create an account with a unique login and password.
   (b) To access paid content, the user's free account must then be connected to the same payment information, which is to be used to maintain that user's access to the specialised content.
   (c) To access the video uploading application, the user must first create a free account, which is then connected to the uploader. This allows the user (aka the digital content creator) to receive monetary bonuses and other incentives for every uploaded video for which the view count reaches above a predetermined threshold view count or subscription rate.

4. Person A is a registered user of MeTube, who has also signed up for access to the video uploading application. Person A is a longtime user who puts a large amount of time and effort into creating custom satirical videos for a subscriber audience which was curated entirely via MeTube.
5. Person B is a registered user of MeTube and has a subscription to access and download specialised content on the hosting platform.

Scene:

Person A, the content creator, creates a unique new video, called Rad video. Person A uploads the video to MeTube, the content host. The video is then accessed, viewed and downloaded by Person B, the content consumer, via the MeTube specialised content subscription service.

In this scenario, there are three relevant parties, or digital content stakeholders, as provided. These three parties—as well as their roles, relationships and connection to the content—have been summarised in Table 2.7.

This example helps to illustrate the relationship between the three primary stakeholders of online user-generated digital content:

1. The content creators
2. The content consumers
3. The content hosts

Table 2.7 MeTube—parties, roles and relationships

Party	Role	Relationship	Connection
Person A	Content creator	• Individual • Registered user of MeTube • Paid uploader of MeTube specialised content	• Creates Rad video • Uploads Rad video to MeTube. • Receives monetary bonuses from MeTube
Person B	Content consumer	• Individual • Registered user of MeTube • Paid subscriber of MeTube specialised content	• Pays subscription fee to MeTube for access to specialised video content • Accesses Rad video on MeTube • Downloads Rad video from MeTube
MeTube	Content host	• Corporation • Commercial online video hosting platform	• Hosts video content platform • Owns and operates primary content access website and video uploading application • Manages and maintains relevant infrastructure • Accepts subscription fee payments • Distributes view count monetary bonuses and royalties

2.5.5 *Digital UCG Stakeholders*

We can group the stakeholders involved in all aspects of online digital user-generated content into three major groups: digital content creators, digital content consumers and digital content hosts.

The **content creators** are the individuals or groups who are the creative developers behind user-generated digital content. It is these users who generate and upload novel content to the online hosting platforms. Highly influential content creators, or those whose content attracts a high number of views, may be rewarded with monetary compensation, royalties or other incentives, provided by the hosting platform as a means of encouraging the content creator to continue producing and uploading novel content. Creators of digital content can include a wide range of platform users as well as artists, performers, musicians, designers, comedians, influencers, consumers and more.

The **content consumers** are the viewers, readers, observers and absorbers of user-generated content. We can subdivide consumers of user-generated content into two consumer subtypes: the freeloader and the purchaser/subscriber.

The **content hosts** are the online platforms—or websites—which host user-generated content. That is, the host provides the uploading application, infrastructure and maintenance, necessary to operate the online hosting platform. This category includes any digital service provider that facilitates interactions between two or more distinct but interdependent sets of users—the creators and the consumers—who interact with each other through the service via the Internet. There are a handful of subtypes of hosting platforms. Examples are provided in Table 2.8.

Table 2.8 Types and examples of digital media hosting platforms

Type of UCG platform	Purposes/functions/benefits	Examples
Social media networks	• Associating individuals and brands • Encouraging individuals and business to interact online and share information to promote mutually productive relationships • Branding, social awareness, relationship building, customer service, lead generation, conversion, expanding range of social influence, etc.	• Facebook • Twitter • LinkedIn
Media-sharing networks	• Finding and sharing photographs, live videos, videos and other media • Brand building, lead generation, targeting, conversion	• YouTube • Instagram • Snapchat • TikTok
Discussion forums	• Finding, sharing and discussing different kinds of information, opinions, news, ideas, etc. • Doing market research with massive reach potential • Running social media marketing campaigns and impactful advertising	• Reddit • Quora • Digg
Content curation networks	• Finding, sharing, discussing and saving a variety of current content and trending media • Channelling brand awareness, generating website traffic and customer engagement and optimising headlines and images for greater accessibility and easier content sharing	• Pinterest • Flipboard
Consumer review networks	• Finding, sharing and reviewing information about a variety of products, services, brands or companies • Generating positive reviews to increase the credibility and reputation of the brand • Resolving customer issues	• Tripadvisor • Yelp • Zomato • Goodreads
Blogging/publishing networks	• Publishing, discovering and commenting on articles, social media blogs and other published written content • Targeting, attracting, engaging and converting a target or niche audience online • Traditional blogging, micro-blogging, social publishing and content marketing	• WordPress • Blogger • Tumblr • Medium • Squarespace
Social shopping networks	• Finding, following, sharing and purchasing. • Creating brand awareness, boosting engagement and selling products without a physical geographical location • Integrating shopping experiences and social experiences	• Etsy • Polyvore • Kijiji • Facebook Marketplace
Sharing economy networks	• Advertising, buying, finding, selling, sharing and trading a variety of products and services among peers	• Uber • Lyft • AirBnB
Anonymous social networks	• Snooping, venting, gossiping and sharing confessions anonymously	• Whisper • Ask.Fm • After school

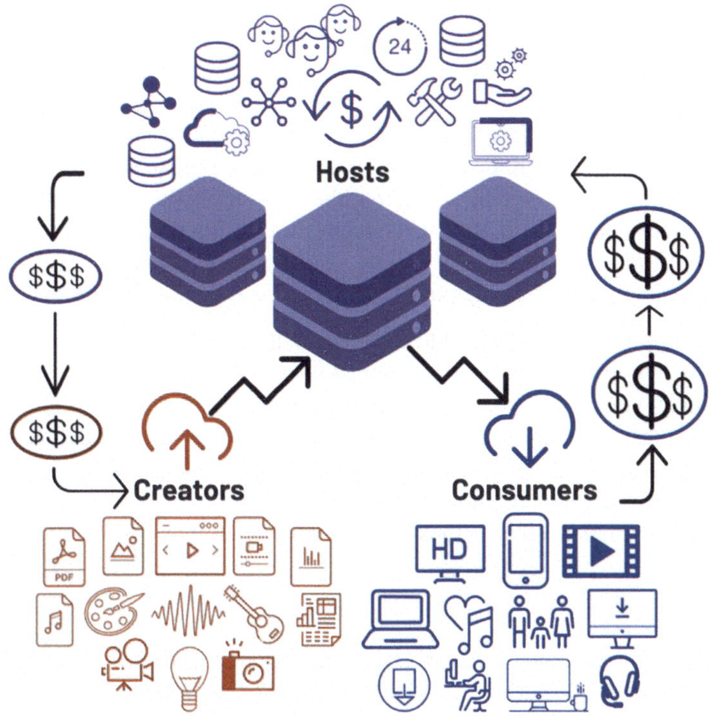

Fig. 2.6 Digital UGC—stakeholder relationships

Arguably, we could also include a fourth stakeholder group: **the communities**, as a tertiary or indirect stakeholder. We will delve into this idea in much greater detail in the next section, when we explore the concept and nature of corporate social responsibility (CSR) as applied through a digital lens.

For additional clarity, Fig. 2.6 below illustrates the roles and relationships between the three main stakeholders of digital user-generated content.

Arguably, we could also include a fourth stakeholder group: the Communities, as a tertiary or indirect stakeholder. We will delve into this idea in much greater detail in the next section, when we explore the concept and nature of corporate social responsibility (CSR) as applied through a digital lens.

2.6 Summary

Digital governance intersects with a wide range of concepts, including ethics, the social contract, social responsibility, sustainability and more. In this chapter, we have explored the concepts of data, first through the lens of ownership and property law and then in terms of data privacy and security. We discussed the fundamentals of data storage and the challenges that can present when dealing with transnational storage infrastructure. Finally, we outlined the concepts of data residency and data localisation both in contrast and in tandem with data sovereignty.

In the next chapter, we will explore the confounding nature of interjurisdictional, transnational and global digital governance relations, particularly with respect to data sovereignty and its implications for governmental organisations, corporations and individual citizens. We will delve into the exciting world of international law, treaties and the issues which arise while navigating conflicting laws between jurisdictional territories.

The answers to the following questions can be found within this chapter:

1. What is digital governance and to whom does it apply?
2. Why might corporations care about social responsibility?
3. Why might consumers care about digital corporate social responsibility?
4. What are three examples of sustainable practices for corporations that operate online?
5. What is a social contract and why do we care?
6. What differentiates user-generated content from other types of digital content?

Additional Sources

- Part I of the Constitution Act, 1982. Canadian Charter of Rights and Freedoms, s 7, Part 1 of the Constitution Act, 1982, being Schedule B to the Canada Act 1982 (UK), 1982, c 11
- Canadian Human Rights Act (RSC, 1985, c H-6)
- United Nations. Universal Declaration of Human Rights. 1948
- Privacy Act (RSC, 1985, c P-21)
- Access to Information Act (RSC, 1985, c A-1)
- Personal Information Protection and Electronic Documents Act (SC 2000, c 5) ("PIPEDA")
- Protecting Canadians from Online Crime Act (SC 2014, c 31)
- Copyright Act (RSC 1985, c C-42)
- Canada's Anti-Spam Legislation (SC 2010, c 23)
- Criminal Code (RSC (1985), c C-46)
- Bill C-10
- Bartsch, R., 2020. The relationship of drug and human trafficking and their facilitation via cyryptomarkets and the dark web: A recommendation for cryptocurrency regulation
- The Palgrave Handbook of International Cybercrime and Cyberdeviance
- Kaur, G., 2021. Understanding Cybersecurity Management in Fintech.
- Martin-Bariteau, F. and Scassa, T., 2021. Artificial Intelligence and the Law in Canada
- Geist, M., 2015. Law, Privacy and Surveillance in Canada in the Post-Snowden Era

- World Health Organization Framework Convention on Tobacco Control (United Nations, *Treaty Series*, vol. 2302, No. 41032)
- Sendai Framework for Disaster Risk Reduction 2015–2030 (resolution 69/283, annex II)
- United Nations Convention on the Law of the Sea (United Nations, Treaty Series, vol. 1833, No. 31363)

Chapter 3
Conflicts of Law

Conflict of laws—also called **private international law**—is the set of rules or laws a jurisdiction applies to a case, transaction, or other occurrences that has connections to more than one jurisdiction. For example, when the parties involved in a dispute are from different legal jurisdictions, when parties from different jurisdictions are engaged in a matter in yet another jurisdiction, or when there are conflicting legal remedies for an issue which arises between legal jurisdictions.

This body of law deals with three broad topics: jurisdiction, rules regarding when it is appropriate for a court to hear such a case; foreign judgments, dealing with the rules by which a court in one jurisdiction mandates compliance with a ruling of a court in another jurisdiction; and choice of law, which addresses the question of which substantive laws will be applied in such a case. These issues can arise in any private-law context, but they are especially prevalent in contract and tort law.

3.1 Topic Overview

In this chapter, we will discuss the concepts of jurisdictionality and how jurisdiction for a legal issue can be determined. We will discuss the personal connecting factors and determinants for jurisdiction to hear an issue. Finally, we will explore international law and related legal strategies in dealing with conflicting laws between jurisdictions.

M. Lukings, A. Habibi Lashkari, *Understanding Cybersecurity Law in Data Sovereignty and Digital Governance*, Progress in IS,
https://doi.org/10.1007/978-3-031-14264-2_3

3.2 Jurisdictionality

Jurisdiction refers to the power, the right or the authority to interpret and apply the law. In legal systems, jurisdictional authority refers to the practical legal authority to make, enforce and administer laws and justice, which is granted to a legal body based on the type and locational circumstances of the case. Jurisdiction can also refer to a specific regional, physical, territorial or geographic area in more casual terms.

When a crime takes place or a dispute occurs between two parties who reside within the same geographic location, the determination of which law applies to that scenario is easily distinguished as being the law of that geographic location. When a dispute occurs between two parties who reside in different nations—or legal jurisdictions—the methodology for determining which country's laws and legal forum should apply is rooted in the topics of private international law, international criminal law and conflicts of law. While these topics are by no means straightforward or simple, the strategy for addressing legal disputes or criminal offences which occur between or across legal jurisdictions becomes greatly complexified when the alleged crime, dispute or other offence takes place online or is otherwise enabled via the Internet.

While jurisdiction is often linked to sovereignty over a territorial location, jurisdiction can also exist without a connection to territory. The type of jurisdictional authority held by a governing body indicates whether that nation or state can undertake enforcement action to uphold its law. This becomes more complex when we consider online activities, particularly cybercrimes, defamation and other tortious activities, as there may not necessarily be a defined geographic area to distinguish which governing authority has the legal jurisdiction to address these matters.

The court of jurisdiction refers to which court has the right or the power to hear a case or argument and render a judgement. It is important, when deciding on a court of jurisdiction, to determine which type of matter the court will hear, as each type of court hears only certain types of arguments or cases. For example, in matters of national law, a federal court would be the appropriate court, as it has legal jurisdiction over matters on a national level. Conversely, minor civil matters may be heard in small claims, civil or municipal courts; family courts handle matters of family law; criminal courts handle criminal matters; and appeal courts handle appeals of previous judgements made in lower courts.

3.2.1 Jurisdictional Authority

There are three main types of jurisdictional authority that grant courts legal power over certain matters: prescriptive jurisdiction, enforcement jurisdiction and adjudicative jurisdiction.

Prescriptive jurisdiction—or legislative jurisdiction—refers to the authority of the governing body of a nation or state to establish laws and legal norms that are applicable to individuals, groups, corporations, property and events, both within and outside of its territory. Under the prescriptive jurisdiction, the laws of a nation or state are still binding on citizens of that jurisdiction while abroad. The same principle of legal scope may also be applicable to certain events or activities conducted abroad that could negatively impact the nation or state which is hoping to assert a prescriptive jurisdiction. For example, a nation may choose to create legislation applicable to crimes that occur abroad which the home nation considers to be a threat to its security or economic interests.

Enforcement jurisdiction refers to the power of a nation or state to ensure compliance with prescriptive legal commands which regulate people and situations in the jurisdiction of that nation or state. Enforcement jurisdiction is closely tied to the adjudicative jurisdiction, and both can be contrasted with the prescriptive jurisdiction.

Adjudicative jurisdiction refers to the power of the governing body of a nation or state to hear and settle legal disputes, as well as the authority to decide and determine the outcomes of competing legal claims. Both the adjudicative jurisdiction and enforcement jurisdiction are territorially limited. The intention for this is to limit the power of a nation or state to enforce its prescriptive or adjudicative jurisdiction within another nation or state. In this way, the legal enforcement and court systems are restricted to operating within the territorial boundaries of their corresponding nation or state. In the absence of permission, a nation or state cannot exercise its prescriptive jurisdiction—either through enforcement or adjudication—outside of its territory.

3.2.2 Determining Jurisdictional Authority

When a legal issue arises between two or more parties and those parties are not bound by the same laws, we call this a "conflict of laws". We see this arise most often with regard to private international law, particularly in the corporate realm.

There are three factors to consider when determining the appropriate court to hear a legal matter, both domestically and on an international level. These correspond with the three types of jurisdictional authority and include jurisdiction of law, recognition of law, and enforcement of law.

1. **Legal Jurisdiction**
   Question: Does the court have the judicial (or legal) power over both the person going on trial and the type of matter that is being tried?
2. **Legal Recognition**
   Question: Will a decision by the court be accepted (or recognised) in foreign countries and/or other jurisdictions?
3. **Legal Enforcement**

Question: Does the court have the power to instil (or enforce) the punishment that is being sought out in this matter?

Domestically, when a corporation exists within a nation and does business between different regional jurisdictions, the courts may look for evidence of the intent of the corporation to do business in one region in contrast to its presence in another. Evidence of intent to do business may be proved through various means, including contractual agreements, targeted sales, advertising campaigns or some types of contact made within another region. Online, such contact could be that which is made by the operators of a website with businesses in a specific geographic area via the availability of the website itself.

Contact, as evidence of an intent to do business, could also be established if the operators made business trips, had telephone calls or exchanged fax messages to the region. As there are many different types of businesses and business engagement, there is a wide variety of other means for proving an intent to do business within a specific region. Based on the subjective nature and wide variety of businesses and business activities, these are often determined by the court of jurisdiction on a case-by-case basis.

Unsurprisingly, the establishment of Internet or online jurisdiction at the international level is not regulated the same way that Internet jurisdiction may be regulated on a provincial, state or domestic regional level. However, the three main features for the determination of jurisdiction must still be considered in this process: judicial (or legal) jurisdiction, enforcement and recognition.

One of the threshold concepts in private international law is the idea that each of us has a "personal law" that goes with us wherever we go, distinct from the law that governs all persons within a given territory. Personal law governs the essential validity of a marriage, the effect of marriage on spousal property rights, jurisdiction in divorce and nullity proceedings and much of the law of succession, among other things.

The idea traces to Roman times, where the word "personal" concerned rulings for a specific person or class of persons—such as a reward or punishment—rather than to all persons. This original meaning (*lex personals*, plural: *leges personals*) gained credence in Montesquieu's 1748 *L'esprit des lois* and from there became enshrined in Article 3 of the *Napoleonic Code*. That article declared that laws relating to the status and capacity of persons govern French persons, even those residing in foreign countries.

The *Code* and its European analogues were one of continental colonialism's exports to much of the world, just as the common law, with its own conception of personal law, was one of the British Empire's exports to the former British colonies and possessions.

Personal law may be based on one's religion, ethnicity or culture. This is so in India, for example, with family law for Hindus, Muslims, Christians and Parsis, except in Goa, though there are perennial efforts at adopting a uniform code for the whole country.

The idea of a law that attaches wherever one goes is also relevant to such things as airplanes, ships, corporations and the legal or "juristic" persons. For example, under the international law of the sea, a ship is registered and given the nationality of the country to which registration has been granted, allowing it to travel internationally.

3.2.3 Personal Connecting Factors

In the secular modern-day world, personal law generally manifests in one of three concepts: nationality, domicile or residence. These three concepts are known as **personal connecting factors**.

To gain facility in private international law, one needs two know something about all three concepts, though we will emphasise domicile—the principal concept in the common law—and residence (the principle concept in the drafting of international conventions, European regulation and statutory reform of the common law).

Personal connecting factors arise in many conflicts analyses. They have diminished relevance in 2020 because people move around a lot, but they still come up in a number of cases.

3.2.3.1 Nationality

Nationality is determined objectively by reference to the legal status of the individual(s) involved in the dispute. It has tended to be deployed as the most appropriate connecting factor in civil law jurisdictions in cases involving real people. Note that "nationality" cannot be used to resolve a CoL dispute within a federalist dispute for obvious reasons and is therefore not generally seen in Quebec. Nationality denotes membership in a nation or sovereign state. It gives a state jurisdiction over the person and gives the person the protection of the state against other states. It may also impose requirements on a person, such as a requirement for military service or a requirement to pay taxes no matter where the person is situated.

It thus has both private-law and public-law aspects and finds expression in the Latin terms *lex patriae* (law of the fatherland) or *lex ligeantiae* (law of the country to which a person owes national allegiance).

Nationality as a personal connecting factor has both advantages and disadvantages. Nationality is related to citizenship, a narrower term often used to denote the status of those nationals who have full political privileges, such as the right to vote. In contrast to domicile, which incorporates subjective intention on where to live out one's days, nationality is based on an objective governmental act. It is thus much easier to ascertain than a person's domicile based on the circumstances of that person's life.

For the same reason, nationality should lead to more uniformity in ascertaining one's personal law. But nationality can persist long after a person severs ties with their patriae, which can make it an artificial and disadvantageous connecting factor.

Also, for countries organised federally, nationality fails to connect a person to the legal system of a unit in the federation. If the legal issue falls under federal jurisdiction, that isn't a problem. But most areas of private law fall under the exclusive, or nearly exclusive, provincial, state or territorial law.

A further complicating factor is that a person can have more than one nationality, giving rise to the question of which should govern—a problem that does not exist with domicile.

3.2.3.2 Domicile and Residence

Domicile, in essence, is residence combined with intention, that is, one objective factor combined with one subjective factor. Factors in which residence is the predominant personal connecting factor are as follows:

- Divorce jurisdiction/recognition of foreign divorces
- Jurisdiction in probate, administration, nullity, adoption, guardianship, mental competency, support, custody and access proceedings
- Recognition of foreign adoptions and of foreign personal representatives
- Effect of marriage on the proprietary rights of spouses
- Formal validity of wills
- Access to no-fault auto insurance schemes

In Canada, the *Charter of Rights and Freedoms* has some effect on the deployment of residency requirements because of the need to protect mobility rights.[1]

What Is Residence?

Broadly, residence is the place where you live. There are many different interpretations of this, but the idea is that it is based on objective criteria about where the person is located and what they're doing there.

Leading Canadian Case: *Thomson v Minister of Natural Revenue*[2]

Residency is "a matter of the degree to which a person in mind and fact settles into or maintains or centralises his ordinary mode of living with its accessories in social relations, interests and conveniences at or in the place in question".

It is possible to have multiple residences at once, unlike domiciles.

---

[1] Part I of the *Constitution Act, 1982*. *Canadian Charter of Rights and Freedoms*, s 7, Part 1 of the *Constitution Act*, 1982, being Schedule B to the *Canada Act 1982* (UK), 1982, c 11.

[2] *Thomson v Minister of National Revenue*, 1946 CanLII 1 (SCC), [1946] SCR 209. Available at: https://canliiconnects.org/en/cases/1946canlii1

Ordinarily Resident

Basically the same as above except with focus on what the person typically does rather than what they're doing right now. Consider a person who has lived abroad for some time but does not plan on permanently relocating there. They cannot be said to be residing in Canada, but they may be ordinarily resident there. Conversely, a person who works in Canada on a temporary basis may be resident in Canada but be ordinarily resident in their home country since that is where they would usually reside.

Habitual Resident

"Somewhere between residence and domicile". It seems to relate to quality of residency; there must be "a degree of settled purpose". It's often conceptualised as being like a domicile with a lessened intention requirement—and no requirement of exclusivity—or as like "ordinarily resident" with a quality component. The influence of duration is disputed, and cases are contradictory on this point.

Habitual Residence of the Child

Canada utilises "the hybrid approach" to determine the habitual residence of the child, as in the case of *Office of the Children's Lawyer v Balev*.[3] This considers both where the parents intend the child to reside and where the child is most acclimatised.

Why Is Residence so Popular?

The focus on the "subjective" element of domicile often produced perverse and nonsensical outcomes; this is perhaps unsurprising when considered considering the evidentiary burden of proving such a high level of intention.

Residence is a low threshold—for example, "not a tourist"—but intention is quite high. It was the preeminent "personal connecting factor" in common law jurisdictions but is being replaced by residency due to the difficulties in ascertaining what domicile means. However, domicile still enjoys considerable use in Quebec and the United States.

Domicile analysis can get quite complicated when there are multiple residences. Always look for clear intentions of either intent to move "permanently" or "indefinitely".[4] Because evidence of such a clear intention is hard to find, the instinct is to revert to Domicile of Origin unless the facts are obvious.

[3] *Office of the Children's Lawyer v Balev*. 2018 SCC 16.

[4] These words mean different things but both have been accepted in the jurisprudence.

Domicile of Origin refers to the place you were born (*Agulian v Cyganik* is the leading case on this and Domicile of Choice, below).[5] The process whereby a Domicile of Choice is abandoned, and the Domicile of Origin is reverted to, is known as the "doctrine of revival" and has been abandoned in some jurisdictions.

Domicile of Choice

The place where you "clearly and unequivocally" choose to live "for an unlimited amount of time". A domicile of choice may override a Domicile of Origin, but when no Domicile of Choice is indicated (or one cannot clearly be ascertained), the Domicile of Origin prevails. This has evidentiary implications, as seen in the case of *IRC v Bullock*,[6] in which the court discussed a hypothetical in which an individual clearly never intends to return to his Domicile of Origin but has not clearly stated his intention as to a Domicile of Choice, so the Domicile of Origin carries the day.

Example: *Foote v Foote Estate*[7]

Choice gets a nuanced interpretation: it means choice unfettered by necessities of work, illness, crimes committed in one jurisdiction, etc. Ergo, a guy who moves back to his Domicile of Origin because he is dying has not necessarily abandoned their Domicile of Choice.

The test for determining abandonment of Domicile of Choice is as follows:

1. Intention to cease residence
2. Acts that execute that intention

Domicile of Choice is a question of mixed fact and law.

Domicile of Children

At common law, children are considered to be domiciled wherever their parents/ guardians are domiciled. In Canada, habitual residence and the best interests of the child are the factors that determine the domicile of the child.

[5] *Agulian v Cyganik [2006] EWCA Civ 129*. Available at: https://www.familylawweek.co.uk/site.aspx?i=ed1657

[6] *IRC v Bullock* [1976] 1 WLR 1178.

[7] *Foote Estate, Re* (2011), 493 AR 354 (CA), 502 WAC 354, 2011 ABCA 1.

Domicile of the Mentally Incapacitated

People who have always been incapacitated are treated like children. People who become incapacitated have their domicile fixed at whatever it was before the incapacitation. Some case law/legal reform on judicial discretion allows an application to change someone's domicile.

Domicile of Corporate Bodies

Countries are domiciled in the jurisdiction they are incorporated in. We can see this in the case of *National Trust Company Limited v Ebro Irrigation and Power Company Limited*[8] among others.[9] In practice, most jurisdictions have rules that require corporations to have presence in the place before they can do business there.[10]

3.2.3.3 Jurisdiction Simpliciter

At common law, courts can generally assume jurisdiction over anyone inside their borders (and not over anyone outside of their borders). In British Columbia, Saskatchewan and Nova Scotia, this rule has been modified by the *Court Jurisdiction and Proceedings Transfer Act.*

Defendants Within the Jurisdiction

If you can reach "em, you can sue em". Somewhat trickier for corporate entities. For corporate defendants, use the "carry on business" for determining jurisdiction in personam since servicing a corporate defendant within the jurisdiction may depend on extrinsic factors that have nothing at all to do with whether or not it's a good idea for the dispute to happen in the jurisdiction.

Defendants Outside the Jurisdiction

Originally at common law, there was no concept of *ex juris* service. That has changed. The general rule post-*Morguard* is that assumption of jurisdiction requires a "real and substantial connection" (for constitutional reasons, provinces are entitled

[8] *National Trust Co. Ltd. v Ebro Irrigation and Power Co. Ltd.* [1954], 3 DLR 326 (Ont.H.C.).

[9] *Gasque v Inland Revenue Commissioners* [1940], 2 KB 80.

[10] *Voyage Co. Industries v Craster* [1998] BCJ No 1884 (Unreported) (BCSC).

to figure out what exactly that means for themselves). This is somewhat supplanted by the *Court Jurisdictions and Proceedings Transfer Act*.

Presumptive Factors for Real and Substantive Connection

Rebuttable presumptive factors, from *Van Breda*:[11]

(a) The defendant is domiciled or resident in the province.
(b) The defendant carries on business in the province.
(c) The tort was committed in the province.
(d) A contract connected with the dispute was made in the province.

As a general rule, courts should not assume jurisdiction in the absence of any connecting factor. Courts do have the discretion to do this, but it should be exercised extremely rarely and only for principled reasons.

On what basis can the presumption be rebutted? The presumption must be rebutted by showing that the connecting factor isn't as useful as it seems. For example, if the connecting factor is a contract, the fact that the contract is completely unrelated to the dispute can rebut the presumption. Similarly, if the defendant carries on business in the province, the presumption of this as a connecting factor can be rebutted by showing that the alleged tort had nothing to do with the defendant's business activities.

A court cannot decline to exercise jurisdiction at this stage if a connecting factor exists and the presumption of jurisdiction is not rebutted except through application of the necessity doctrine. This does not preclude a court declining to exercise jurisdiction if a court raises the issue of *forum non conveniens* being raised by one of the parties.

In a multi-tort case, if the real and substantial connection is made out on one tort, the court assumes jurisdiction over all related tort actions. Expecting a plaintiff to pursue the claim in multiple jurisdictions after finding a real and substantial connection would be "incompatible with any notion of fairness and efficiency".

There is also a test for identifying new connecting factors (also from *Van Breda*):

(a) Similarity of the connecting factor with the recognised presumptive connecting factors
(b) Treatment of the connecting factor in the case law
(c) Treatment of the connecting factor in statute law
(d) Treatment of the connecting factor in the private international law of other legal systems with a shared commitment to order, fairness and comity

[11] *Club Resorts Ltd. v Van Breda*, 2012 SCC 17 [2012] 1 SCR 572.

Example: *Moran v Pyle National (Canada)*[12]

The plaintiff is the family of a dead electrician killed while removing a defendant-manufactured light bulb while in Saskatchewan (the jurisdiction of the forum). The defendant manufactures light bulbs sold in Saskatchewan but otherwise has no presence in the province. When you manufacture something negligently and it's reasonably foreseeable that it will make it into the jurisdiction where the Court of Appeal takes place, that's enough to get you to real and substantial connection.

Example: *Court v Debaie*[13]

The defendant makes defamatory comments on the plaintiff's Facebook page. The plaintiff wants to sue in Alberta because some of the people who saw the defamatory comments were in Alberta. This is sufficient to count as "published" in Alberta and therefore that the tort was committed in Alberta. This case also has some commentary on the admissibility of inadmissible evidence on threshold questions like jurisdiction.

Example: *Tamminga v Tamminga* [2014 ONCA 478]

The plaintiff gets in an accident and sues insurance company in Ontario even though the accident happened elsewhere on the basis that an insurance contract was signed in Ontario. In this case, it was held that this contract is not sufficiently connected to the dispute to entitle the court to take jurisdiction.

Class Actions

Jurisdiction simpliciter gets notably more complex in a class action: Consider a situation where the real and substantial connection is based on the tort being committed in the province. How can non-resident members of the class make out a real and substantial connection to the forum?

3.2.4 Subtypes of Jurisdiction

The term jurisdiction is derived from the two Latin terms *iuris* meaning "law" and *dicere* meaning "to speak". Jurisdiction is therefore the authority granted to a

[12] *Moran v Pyle National (Canada) Ltd.* [1975] 1 SCR 393.

[13] *Court* et al. *v Debaie*, 2012 ABQB 640.

formally constituted legal body to deal with and make pronouncements on legal matters. Jurisdiction is therefore the authority to administer justice within a defined area of responsibility. Jurisdiction also denotes the geographical area or subject matter to which such authority applies.

3.2.4.1 Juridical Jurisdiction

There are three main types of judicial jurisdiction: personal, territorial and subject matter:

- **Personal Jurisdiction**—the authority over a person, regardless of their location
- **Territorial Jurisdiction**—the authority confined to a bounded space, including all those present therein, and events which occur there
- **Subject-Matter Jurisdiction**—the authority over the subject of the legal questions involved in the case

When the court exercises jurisdiction over property located within the perimeter of its powers without regard to personal jurisdiction over the litigants, it is called jurisdiction in rem.

The subject-matter jurisdiction of certain courts is limited to certain types of controversies (e.g. suits in admiralty or suits where the monetary amount sought is less than a specified sum). These are referred to as courts of special jurisdiction or court of limited jurisdiction. A court whose subject matter is not limited to certain types of controversy is referred to as a court of general jurisdiction.

Original jurisdiction and appellate jurisdiction are also other types of jurisdictions. A court of original jurisdiction hears cases as they are first initiated by a plaintiff, but a court of appellate jurisdiction may only hear an action after the court of original jurisdiction (or a lower appellate court) has heard the matter.

3.2.4.2 Personal vs Territorial vs Subject-Matter Jurisdiction

Personal jurisdiction refers to the jurisdiction of the person, the party in the case.

Territorial jurisdiction refers to the geographical area within which a court may exercise its powers. With territorial jurisdiction, a regional or district court has jurisdiction within the region or district in which it is situated, and a court of a resident magistrate has jurisdiction in the region within which it is situated.

Subject-matter jurisdiction refers to the authority of a court to hear cases of a particular type or cases relating to a specific subject matter. For instance, bankruptcy courts only have the authority to hear bankruptcy cases. Subject-matter jurisdiction is the requirement that a given court have power to hear the specific kind of claim that is brought to that court. While litigating parties may waive personal jurisdiction, they cannot waive subject-matter jurisdiction.

For example, subject-matter jurisdiction means that family matters, such as divorce and child custody, can only be heard by a court specialising in family law.

Similarly, criminal cases would not be heard by a family court, but by a court specialising in criminal law.

3.2.4.3 Exclusive vs Concurrent Jurisdiction

Courts may also have jurisdiction that is exclusive or concurrent (shared). Where a court has exclusive jurisdiction over a territory or a subject matter, it is the only court that is authorised to address that matter. Where a court has concurrent or shared jurisdiction, more than one court can adjudicate the matter. In cases where concurrent jurisdiction exists, a party may attempt to engage in forum shopping, by bringing the case to a court which the party presumes will rule in its favour.

Exclusive jurisdiction means that an action can be commenced only in the chosen forum. Non-exclusive jurisdiction provides some flexibility to the parties as it authorises more than one jurisdiction to hear the action. Parties intending exclusive jurisdiction must clearly indicate it in the clause. Exclusive jurisdiction exists in civil procedure if one court has the power to adjudicate a case to the exclusion of all other courts. The opposite situation is concurrent jurisdiction in which more than one court may take jurisdiction over the case. Exclusive jurisdiction means that only a particular court can decide a case.

For example, bankruptcy court is a court with exclusive subject-matter jurisdiction.

Concurrent jurisdiction exists where two or more courts from different systems simultaneously have jurisdiction over a specific case. This situation leads to forum shopping, as parties will try to have their civil or criminal case heard in the court that they perceive will be most favourable to them.

For example, in the United States federal and state courts may have concurrent jurisdiction over specific crimes. For example, a person who robs a bank may be tried and convicted in state court for robbery and then tried and convicted in federal court for the federal offence of robbery of a federally chartered savings institution.

3.2.4.4 Original vs Appellate Jurisdiction

Original jurisdiction is the right of a court to hear a case for the first time. It can be distinguished from appellate jurisdiction which is the right of a court to review a case that has already been heard and decided upon by a lower court.

Appellate jurisdiction refers to the power of a court to hear appeals from lower courts. An appellate court, commonly called a court of appeal, appeal court, court of second instance or second instance court, is any court of law that is empowered to hear an appeal of a trial court or other lower tribunal.

3.2.4.5 General vs Limited/Special Jurisdiction

General jurisdiction refers to a court's competence to adjudicate disputes arising out of a defendant's activities anywhere in the world. A court of general jurisdiction is a court with authority to hear cases of all kinds—criminal, civil, family, probate and so forth.

Courts of general jurisdiction are granted authority to hear and decide all issues that are brought before them. These are courts that normally hear all major civil or criminal cases. For example, courts of general jurisdiction will hear:

- **Felonies—**serious crimes including assault, murder and crimes committed with a deadly weapon
- **Misdemeanours—**slightly less serious crime, such as first-time drug offences, assault without a deadly weapon, etc.
- Fraud, bankruptcy and other so-called "white collar" crimes

Limited jurisdiction, or special jurisdiction, is the court's jurisdiction only on certain types of cases such as bankruptcy, claims against the government, traffic, probate, family matters, immigration and customs or limitations on courts' authority to try cases involving maximum amounts of money or value.

Courts of limited jurisdiction, as opposed to general jurisdiction, derive power from an issuing authority, such as a constitution or a statute. These "special courts" are those courts with a limited jurisdiction—such as traffic court, probate court, family courts, military courts, misdemeanour courts, domestic violence court or small claims court, to name a few. Special courts also include courts which serve a particular population, such as tribal courts.

3.2.4.6 Diversity Jurisdiction

Diversity jurisdiction is a form of subject-matter jurisdiction that arises when there are multiple plaintiffs and multiple defendants and no plaintiff is from the same state as any defendant.

For example, in the law of the United States, diversity jurisdiction gives US federal courts the power to hear lawsuits that do not involve a federal question. Diversity jurisdiction requires two conditions to be met: first, there must be "diversity of citizenship" between a lawsuit's parties, meaning the plaintiffs must be citizens of different US states than the defendants and, second, the lawsuit's "amount in controversy" must be more than $75,000. If a lawsuit does not meet these conditions, US federal courts will normally lack the power to hear it unless it involves a federal question, and the lawsuit would need to be heard in state court instead. As an example, if plaintiffs from Texas, Georgia and Illinois jointly sue three defendants from Missouri, Maine and New Jersey, there is diversity jurisdiction.

3.2.4.7 Pecuniary Jurisdiction

Pecuniary[14] **jurisdiction** tries to address whether a court of law can try cases and suits of the monetary value/amount of the case or suit in question. For example, consumer courts have different pecuniary jurisdictions. In determining whether there is diversity jurisdiction, a corporation that is a party is considered to be a citizen of both its state of incorporation and its principal place of business.

3.2.4.8 Example: United States

In some countries, including the United States, jurisdiction is conceptually divided between jurisdiction over the subject matter of a case (subject-matter jurisdiction) and jurisdiction over the person of the litigants (personal jurisdiction). In the United States, all states have courts of general jurisdiction; most states also have some courts of limited jurisdiction. For example, federal courts in the United States are courts of limited jurisdiction.

Federal jurisdiction can be divided into two as:

1. Federal question jurisdiction
2. Diversity jurisdiction

The US district courts also have jurisdiction to hear only cases arising under federal law and treaties, cases involving ambassadors, admiralty cases, controversies between states or between a state and citizens of another state, lawsuits involving citizens of different states, and against foreign states and citizens.

Certain courts, particularly the US Supreme Court and most state supreme courts, have discretionary jurisdiction. By virtue of this, these courts can choose which cases to hear from among all the cases presented on appeal. Such courts generally only choose to hear cases that would settle important and controversial points of law. The most important aspect of discretionary jurisdiction is that though these courts have discretion to deny cases they can adjudicate; no court has the discretion to hear a case that falls outside of its subject-matter jurisdiction.

For example, the US district courts have original jurisdiction over a number of different matters, but the US Court of Appeals has appellate jurisdiction only over matters appealed from the district courts. In the case of *Georgia v South Carolina*, it was held that the Supreme Court had original jurisdiction in a case involving the correct location of a boundary between the two states.

However, in a special class of cases, the US Supreme Court has the power to exercise original jurisdiction. Under 28 USC. § 1251, the Supreme Court has original and exclusive jurisdiction over controversies between two or more states and original (but non-exclusive) jurisdiction over cases involving officials of foreign

[14] "Pecuniary" means "related to money".

states, controversies between the federal government and a state and actions by a state against the citizens of another state or foreign country.

3.3 International Law

In law, we can distinguish (for the most part) between international and domestic law, as well as between public and private law. International law involves the regulation of relationships between sovereign states, while domestic law, on the other hand, confers rights to persons and entities within those individual sovereign states. Put simply, international law refers to the rules which regulate the relations between various nations of the world. Domestic law and international law are enforced differently, primarily because international law deals with the issue of sovereignty. International law refers to the rules which regulate the relations between various nations of the world.

3.3.1 Sources of International Law

Sovereignty, in international law, refers to the legal power and authority of a governing body to exercise control over a nation or state and the autonomy of that governing body to do so freely and without external pressure, control or influence.

In international law, sovereignty is the exercise of power by a state. *De jure* sovereignty refers to the legal right of the state to exercise their power and authority. This can be contrasted with de facto sovereignty, which refers to the actual factual ability of a state to exercise their power and authority rather than the legal right to do so. While we typically expect *de jure* and *de facto* sovereignty to exist at the same place, at the same time and within the same governing organisation, a failure of that expectation can become an issue of concern for the determination of sovereignty in international law. Although the twentieth century resulted in greater limitations on sovereignty, it continues to be a major issue in international law, particularly in cases of international human rights violations and regional genocide.

International considerations include legal documents and formalised contractual agreements between nations, known as treaties, as well as other international sources of influence on domestic policymaking. In this section, we will discuss the relevant treaties and international agreements related to cybersecurity and cyber law, the creation of the *Tallinn Manual* as a response to expressed international concern over cyber warfare and the international legal principles as seen in international humanitarian law and private international law which influence national decision-making.

3.3.1.1 Treaties and Conventions

The United Nations (UN) is an intergovernmental organisation aiming to maintain international peace and security, develop friendly relations among nations, achieve international cooperation and be a centre for harmonising the actions of nations. One of the legal forms of cooperation used by the United Nations is the signing and implementation of treaties between individual nations.

The Council of Europe is one of several regional organisations established in the aftermath of World War II. It is separate and distinct from the European Union and has a much larger membership than the European Union. The Council of Europe's core mission is the protection of human rights, but it also works to promote democracy, the rule of law and uniform standards. Much of the Council's work is accomplished through the drafting of treaties.

In late 2009, the Cooperative Cyber Defence Centre of Excellence convened an international group of legal scholars and practitioners to draft a manual addressing the issue of how to interpret international law in the context of cyber operations and cyber warfare and to bring some degree of clarity to the associated complex legal issues. The *Tallinn Manual* was produced between 2009 and 2012 by an international team of legal scholars at the request of the North Atlantic Treaty Organization (NATO) Cooperative Cyber Defence Centre of Excellence, which is located in Tallinn, Estonia. The *Tallinn Manual on the International Law Applicable to Cyber Warfare* (more simply known as the *Tallinn Manual*) was published by Cambridge University Press in April 2013.[15] The *Manual* was the first comprehensive and authoritative attempt to analyse the application of international law to cyber warfare.

As an academic and non-binding study, the *Tallinn Manual* followed similar efforts, such as the *San Remo Manual on International Law Applicable to Armed Conflicts at Sea*[16] by the International Institute of Humanitarian Law and the *Manual on International Law Applicable to Air and Missile Warfare*,[17] as written by the Harvard Program on Humanitarian Policy and Conflict Research. While it was produced upon invitation from a NATO organisation, the *Tallinn Manual* is not a NATO document, rather it is an independent academic research product representing only the views of its authors in their personal capacity. The manual does not represent the views of NATO nor any other organisation or state, including those represented by the observers.

---

[15] Schmitt, M. (2013). *Tallinn Manual on the International Law Applicable to Cyber Warfare*. Cambridge: Cambridge University Press. doi:10.1017/CBO9781139169288.

[16] International Institute of Humanitarian Law (1995). *San Remo Manual on International Law Applicable to Armed Conflicts at Sea* (L. Doswald-Beck, Ed.). Cambridge: Cambridge University Press. doi:10.1017/CBO9780511622052.

[17] Program on Humanitarian Policy and Conflict Research at Harvard University (2013). *HPCR Manual on International Law Applicable to Air and Missile Warfare*. Cambridge: Cambridge University Press. doi:10.1017/CBO9781139525275.

The manual is divided into sections of rules and their accompanying commentary. The rules themselves are, in essence, restatements of international law in the cyber context, as understood and agreed to, by all of the authors. Being the first authoritative restatement of the application and interpretation of international law in the cyber context, however, it may be anticipated that the manual will have an effect on how states and organisations will formulate their approaches and positions in those matters.

In February 2017, a follow-up report, the *Tallinn Manual 2.0 on International Law Applicable to Cyber Operations*, was released.[18] *The Tallinn Manual 2.0* broadens the scope to assess how international legal principles can be applied to malevolent cyber operations that do not rise to the level of an armed attack. The focus of the original *Tallinn Manual* is on the most disruptive and destructive cyber operations—those that qualify as "armed attacks" and therefore allowing states to respond in self-defence and those taking place during armed conflict.[19] Since the threat of cyber operations with such consequences is especially alarming to states, most academic research has focused on these issues. *Tallinn 2.0* refers to "cyber operations" as opposed to "cyber conflicts" as in the original *Tallinn Manual.*[20,21]

3.3.1.2 The Rule of Law

The **rule of law** is a legal principle which suggests that every person is subject to the law, including people who are lawmakers, law enforcement officials and judges. The principle of the rule of law goes back to ancient philosophers, including Plato and Aristotle.

While the modern use of the phrase "the rule of law" has been credited to Samuel Rutherford, John Locke and A. V. Dicey, the development of the fundamental legal concept of the rule of law can be traced back through history and linked to many ancient civilisations as far back as ancient Greece, Mesopotamia, India and Rome. Plato, an Athenian philosopher in the Classical period of ancient Greece, advocated for a benevolent monarchy ruled by an idealised philosopher king. While the philosopher king would be above the law, Plato expressed the ideal that the best men of the benevolent monarchy would be good at respecting existing laws.

In comparison to Plato, Aristotle was more strongly opposed to letting the highest officials wield power beyond guarding and serving the laws, what we now recognise as the legal principle of the rule of law. In jurisprudence, the "rule of the best"

[18] Schmitt, M. (2017). *Tallinn Manual 2.0 on the International Law Applicable to Cyber Operations* (second ed.). Cambridge: Cambridge University Press. doi:10.1017/9781316822524.

[19] Schmitt, M. (2013). *Tallinn Manual on the International Law Applicable to Cyber Warfare*. Cambridge: Cambridge University Press. doi:10.1017/CBO9781139169288.

[20] Schmitt, M. (2013). *Tallinn Manual on the International Law Applicable to Cyber Warfare*. Cambridge: Cambridge University Press. doi:10.1017/CBO9781139169288.

[21] Schmitt, M. (2017). *Tallinn Manual 2.0 on the International Law Applicable to Cyber Operations* (second ed.). Cambridge: Cambridge University Press. doi:10.1017/9781316822524.

described a system of governance where only the best of the citizens, chosen through a careful process of selection, would become rulers. In this system, hereditary rule would be forbidden, unless the children of the incumbent ruler performed best and were better endowed with the attributes that make a person fit to rule in comparison to every other citizen in the polity. The principle of the rule of law is an important consideration in international law to this day.

3.3.1.3 Judicial Decisions and Teachings

The International Court of Justice (ICJ)—also known as the World Court—is the principal judicial organ of the United Nations, having been established in June 1945 by the *Charter of the United Nations*. Article 38(1) of the *Statute of the International Court of Justice* lists four sources of international law: treaties and conventions, custom, general principles of law and judicial decisions and teachings.

The election of the first members of the International Court of Justice took place at the First Session of the United Nations General Assembly and Security Council on February 1946, with work in the court beginning in April 1946. The Court is composed of fifteen judges, who are elected for terms of office of 9 years by the United Nations General Assembly and the Security Council.

The International Court of Justice only hears lawsuits between nation-state,s and its jurisdiction is not compulsory, meaning that both nations in a dispute must agree to have the International Court of Justice hear the dispute. The seat of the court is at the Peace Palace in The Hague. Of the six principal organs of the United Nations, it is the only one that is not located in New York, in the United States. The role of the International Court of Justice is to settle, in accordance with international law, legal disputes submitted to it by nation-states and to give advisory opinions on legal questions referred to it by authorised United Nations organs and specialised agencies.

3.3.2 *Key Determinants in International Law*

Prescriptive jurisdiction—or legislative jurisdiction—refers to the authority of the governing body of a nation or state to establish laws and legal norms that are applicable to individuals, groups, corporations, property, and events, both within and outside of its territory. Under the prescriptive jurisdiction, the laws of a nation or state are still binding on citizens of that jurisdiction while abroad. The same principle of legal scope may also be applicable to certain events or activities conducted abroad that could negatively impact the nation or state which is hoping to assert a prescriptive jurisdiction. For example, a nation may choose to create legislation applicable to crimes that occur abroad which the home nation considers to be a threat to its security or economic interests.

Enforcement jurisdiction refers to the power of a nation or state to ensure compliance with prescriptive legal commands which regulate people and situations in the jurisdiction of that nation or state. Enforcement jurisdiction is closely tied to the adjudicative jurisdiction, and both can be contrasted with the prescriptive jurisdiction.

Adjudicative jurisdiction refers to the power of the governing body of a nation or state to hear and settle legal disputes, as well as the authority to decide and determine the outcomes of competing legal claims. Both the adjudicative jurisdiction and enforcement jurisdiction are territorially limited. The intention for this is to limit the power of a nation or state to enforce its prescriptive or adjudicative jurisdiction within another nation or state. In this way, the legal enforcement and court systems are restricted to operating within the territorial boundaries of their corresponding nation or state. In the absence of permission, a nation or state cannot exercise its prescriptive jurisdiction—either through enforcement or adjudication—outside of its territory.

3.3.3 International Human Rights Laws

Human rights is a body of law at both the domestic and international levels, which aims to establish a set of civil, political, economic, social and cultural rights to which all people are entitled by virtue of their humanity, inherently.

3.3.3.1 Domestic Law

Every nation and/or state has its own laws which apply to corporations operating under their laws. For example, in Canada, laws relevant to human rights range from those provided in the *Charter of Rights and Freedoms* and comparative constitutional law to provincial and territorial human rights codes.

3.3.3.2 International Treaties

Relevant treaties for corporations which operate on- or offline include the *Universal Declaration of Human Rights,*[22] the *International Covenant on Civil and Political Rights,*[23] the *Convention on the Rights of the Child,*[24] the *United Nations Convention Against Transnational Organized Crime,*[25] the *Optional Protocol to the*

[22] The United Nations (1948). Universal Declaration of Human Rights.

[23] 999 UNTS 171.

[24] https://www.ohchr.org/en/instruments-mechanisms/instruments/convention-rights-child

[25] 2225 UNTS 209.

Convention on the Rights of the Child[26] and the *United Nations Convention on the Use of Electronic Communications in International Contracts.*[27]

The Council of Europe also has a treaty, the *Convention for the Protection of Human Rights and Fundamental Freedoms*,[28] also known as the *European Convention on Human Rights*. The European Court of Human Rights is responsible for monitoring compliance with the *Convention.*[29]

The *Universal Declaration of Human Rights* (UDHR) was proclaimed by the UN General Assembly in 1948.[30] Although the *Declaration* is not legally binding, many of its principles have been incorporated in international treaties, regional human rights instruments and national constitutions.[31]

Article 34 of the *Convention on the Rights of the Child* obligates state parties to protect children from all forms of sexual exploitation and abuse, including prostitution and pornography.[32]

Under international law, states have the legal duty to protect human rights against violations by state and non-state actors and the responsibility to provide appropriate remedies when those rights have been violated. These legal duties are expressed in treaties as well as in the general comments which accompany the treaties and issued periodically by the treaty bodies which monitor the implementation of those treaties by governments. Included within this group of treaties are the ones provided above as well as the *International Covenant on Civil and Political Rights*[33] and the *International Covenant on Economic, Social, and Cultural Rights.*[34] When combined with the *Universal Declaration of Human Rights*,[35] these collectively form what has come to be known today as the *International Bill of Human Rights.*[36]

Unfortunately, the provisions of these treaties are not directly enforceable in courts in Canada, although Canada is bound by them, having ratified these, as treaties require legislative implementation to have a direct effect in Canada, as in other common law countries. However, non-binding human rights law, like the *Universal Declaration*,[37] can eventually become customary international law over time and can then be directly incorporated into Canadian law.

---

[26] 2171 UNTS 227.

[27] https://uncitral.un.org/en/texts/ecommerce/conventions/electronic_communications

[28] https://www.echr.coe.int/documents/convention_eng.pdf

[29] ETS 5.

[30] The United Nations (1948). Universal Declaration of Human Rights.

[31] The United Nations (1948). Universal Declaration of Human Rights.

[32] 1577 UNTS 3.

[33] 999 UNTS 171.

[34] https://www.ohchr.org/en/instruments-mechanisms/instruments/international-covenant-economic-social-and-cultural-rights

[35] https://www.un.org/en/about-us/universal-declaration-of-human-rights

[36] https://www.ohchr.org/en/what-are-human-rights/international-bill-human-rights

[37] https://www.un.org/en/about-us/universal-declaration-of-human-rights

While natural and legal persons are not generally bound by international human rights law, there has been a growing recognition that businesses have a responsibility to protect human rights, as expressed in the *United Nations Guiding Principles on Business and Human Rights*.[38]

3.4 International Legal Strategies

International law structures relations among states and other international stakeholders (most notably international organisations) through various prohibitions, requirements and permissions. As such, it has provided a path for regulating global governance issues from arms control to trade to the environment. As states give increased attention to the governance of cyberspace (the technical architecture that allows the global Internet to function) and governance in cyberspace (how states, industry and users may use this technology), the role of international law in the cyber context has gained increasing prominence.

With few exceptions (most notably, the *Budapest Convention on Cybercrime*[39] and the not-yet-in-force *African Union Convention on Cyber Security and Personal Data Protection*[40]), international law does not have tailor-made rules for regulating cyberspace. Moreover, the technology is both novel and dynamic. Thus, for several years, there were open questions about whether existing international law applied to cyberspace at all. Today, most states and several international organisations, including the UN General Assembly's First Committee on Disarmament and International Security, the G20, the European Union, ASEAN and the OAS, have affirmed that existing international law applies to the use of information and communication technologies (ICTs) by states. As such, the current discourse centres not on whether international law applies but rather how it does so.

Unlike many other international issues, the governance of cyberspace did not originate with states, but with the academic institutions and private actors who constructed the Internet (albeit with government funding). With the Internet's commercialisation, ICT companies emerged; today, their platforms serve as the environment for the vast majority of cyber behaviour, including state and state-sponsored cyber operations. States' interest in cyberspace, particularly as a zone for geopolitical rivalries, followed. Thus, cyberspace governance involves key stakeholders that include, but are by no means limited to, states.

International law, however, is primarily a legal order for states (and their creations, like international organisations). As such, international law does not hold a monopoly on the regulation of cyberspace. Given industry and civil society players,

[38] https://www.ohchr.org/sites/default/files/documents/publications/guidingprinciplesbusinesshr_en.pdf

[39] https://www.coe.int/en/web/cybercrime/the-budapest-convention

[40] https://au.int/en/treaties/african-union-convention-cyber-security-and-personal-data-protection

other regulatory regimes (e.g. industry self-regulation) offer alternative vehicles. Multi-stakeholder governance, for example, has become the main avenue for governance of the Internet's architecture. At the same time, non-state actors have expressed an interest in questions of how international law applies to governance in cyberspace. For example, the International Committee of the Red Cross and the Independent Groups of Experts who authored the two *Tallinn Manuals* (with plans for a third in the works) have explored how existing international law rules and principles translate into the cyber context. Microsoft's President Brad Smith even called on states to conclude a new *Digital Geneva Convention* to regulate state behaviour in cyberspace.[41]

Issues surrounding international law's application to cyberspace may be broken into five discrete categories: (1) silence, (2) existential disagreements, (3) interpretative challenges, (4) attribution and (5) accountability.

3.4.1 Silence

Without tailored-made treaties on cyber issues, the application of international law depends on identifying customary international law rules—that is, state practice accepted as law. For many years, figuring what states were doing in cyberspace—let alone what they thought international law had to say about it—was complicated by state silence.

Over the last decade, however, several states have begun to speak out. Beginning in 2012, the United States started to offer its views in a series of speeches and statements. In 2018, the United Kingdom's Attorney General made an important statement of UK views. In the ensuing years, other (mostly European) states began to offer their own detailed perspectives, including Australia, Estonia, Finland, France, Germany and the Netherlands. Ongoing efforts at the United Nations and in regional contexts like the OAS are now seeking to expand the list of states with opinions on international law in cyberspace. As yet, however, the vast majority of states remain silent.

Why these states remain silent is often unclear. Some may wish to avoid becoming entangled in international disputes among those states who have spoken out. For others, however, the issue may be a matter of legal capacity. Many states lack the personnel or resources to understand the issues involved in applying international law to cyberspace. Thus, a threshold issue for international law's application is building up sufficient legal capacity for all states to have a voice in shaping what international law says on cyber issues.

[41] https://carnegieendowment.org/2021/06/14/brief-primer-on-international-law-and-cyberspace-pub-84763

3.4.2 Existential Disagreements

Among those states that have taken positions on international law's application to cyberspace, there are a number of "existential" disagreements—competing claims that a particular international legal rule or regime is entirely included or excluded from cyberspace. In the United Nations context, for example, a few states have challenged the availability of international humanitarian law, the right of self-defence, the duty of due diligence and the right to take countermeasures with respect to online activity. The existence (or absence) of one or more of these legal frameworks from cyberspace has significant implications for international law's application, impacting how states conduct their cyber operations in armed conflicts, their ability to respond to malicious cyber activity conducted by other states and what actions they must take to protect the rights of third states from harms originating in their own territories.

3.4.3 Interpretive Questions

Even where states accept that a particular international legal rule or regime applies in cyberspace, substantial interpretative questions often remain open to debate. International legal regimes like non-intervention, sovereignty and human rights encounter much ambiguity in their applications to cyberspace. The duty of non-intervention, for example, protects a states' international and external affairs from "coercive" intervention by other states. Yet, there's no consensus on which "affairs" the duty protects, let alone what differentiates coercive from noncoercive cyber activity. Similarly, sovereignty is undoubtedly one of the core architectural features of the international legal order. States appear to diverge, however, on whether sovereignty merely is a foundational principle on which other international legal rules (like non-intervention) rest or if it is an independent rule that can be breached by certain foreign state cyber operations directly.

3.4.4 Attribution

International law only regulates its subjects of international law (e.g. states). It does not usually direct the behaviour of ICT companies or individuals (who are usually subject to one or more domestic legal orders). To apply international law in cyberspace, therefore, it is necessary to know the identity of whoever is responsible for the activity in question: is it a state or state-sponsored actor subject to international law or is it an individual(s) engaged in behaviour outside international law's ambit? Such identifications are, however, difficult in cyberspace given well-known challenges in technical attribution—identifying the origins of malicious cyber behaviour is often difficult and time-consuming. Moreover, where states employ proxies, attribution is further complicated by the need to show evidence of state "control" over the proxy actor (international law has yet to fully resolve how much control is required or what evidence must be shown to demonstrate it).

3.4.5 Accountability

Although attributions of state and state-sponsored cyber operations may be on the rise, accountability has proved challenging. States that accuse other states of malicious cyber behaviour rarely invoke international law in doing so. The absence of international legal rhetoric may imply that the behaviour may be lawful, even if unwanted. Nor has the naming and shaming that has occurred done much to change the accused's behaviour (it may, however, help clarify the existence and meaning of international law in cyberspace). Some states have been working to create coalitions that can cooperate on improving accountability through collective accusations or even sanctions. So far, however, such efforts have not focused on using international law's benchmarks for measuring malicious cyber behaviour nor its tools (e.g. countermeasures) for addressing its violation.

3.5 Interjurisdictional Sovereignty and Governance

Going forward, there are two dominant questions about the application of international law to cyberspace:

(a) What form should any resolution of the main issues take?
(b) In what form should this occur?

The future form(s) of international law? Even as international law lacks tailor-made rules, various states and stakeholders suggest that existing international law is sufficient to regulate behaviour of states: For these actors, the future must address how to operationalise and improve accountability under the current law.

In contrast, other states and stakeholders have suggested that there are gaps or inefficiencies in the existing law that require the formulation of new rules. This could occur via a new treaty or the evolution of customary international law. The proponents of new rules are not, however, aligned on the desired end state. Some states are focused on a treaty that would protect states from people (e.g. those who would engage in subversive speech or otherwise threaten the security of the state through online activity). Others seek a treaty that would protect people from states (e.g. prohibiting states from engaging in malicious cyber operations against critical cyber infrastructure and requiring positive efforts from states to cooperate and forestall such operations by others).

The future forum for applying international law—To date, the most robust fora for addressing international law's application are non-state oriented. Governance of the Internet involves a multi-stakeholder process. And non-state actors like the Independent Groups of Experts who crafted the Tallinn manuals have dominated the discourse on how international law regulates state cyber operations. Such efforts may continue to hold sway, as witnessed by the three recent Oxford statements produced by international lawyers on the law's prohibitions and protections vis-à-vis the healthcare sector and vaccine research amid the COVID-19 pandemic as well as foreign election interference prior to the 2020 US presidential election. At the same time, United Nation member states' efforts to address international law as it applies to cyber/ICT-related issues have multiplied.

Moving beyond the original United Nations Group of Governmental Experts forum, UN processes now also encompass an Open-Ended Working Group in the UN General Assembly's First Committee, as well as a Third Committee process on a United Nations cybercrime convention. It is not, however, a foregone conclusion that the United Nations will be the only landing place for shaping discussions on how international law applies. Regional organisations (such as the European Union) may offer an alternative that in some cases can avoid certain aspects of geopolitics that dominate the UN discourse. Other existing or future multi-stakeholder processes might also take up the mantle.

In short, even as states move past their initial reticence to address international law's application in (and to) the behaviours of different actors in cyberspace, the consequences of new voices in various forms and fora warrant heightened attention. States and other stakeholders should carefully track various issues and players involved with an eye on the big picture—seeing if international law can do more than simply apply in cyber contexts. The real question remains—can it do so in ways that are effective at regulating the behaviour of states and other actors of international concern?

3.5.1 Internet Jurisdiction

Internet jurisdiction is used to determine which legal authority may hear a case, between a defendant and plaintiff, in which the alleged offence or tortious action was committed on the Internet. While in typical matters the plaintiff appeals to the court or legal authority which has jurisdiction over the geographical area in which the crime occurred, this is made much more complex when the issues at hand are committed, remotely, over the Internet. In these cases, it may be difficult to determine an appropriate geographically based jurisdiction, particularly when the originator of the illegal, tortious or otherwise offensive action can be located in a different region or country than the individual or entity against whom the crime or offence was allegedly committed.

Usually, in cases of this nature, each country will take its own laws into consideration when determining a legal action, before considering the laws of the country in which the defendant may live. If one nation determines a crime was committed on the Internet, according to its laws, and determines that its courts have jurisdiction over the case, then that country may prosecute the offender(s), even when those involved may live in a different country.

3.5.1.1 Example: People's Republic of China

We can see an example of the complexities which arise with the assignment of jurisdiction related to cybercrime in the codified *Chinese Criminal Law* provisions, especially Articles 6, 7, 8 and 9.[42,43] Specifically, Article 6 stipulates that "Any person who commits a crime in the territory of the People's Republic of China shall apply this law unless otherwise specifically provided by law".[44]

Based on the established criminal law provisions in China, Article 6 is interpreted to mean that Chinese courts have jurisdiction over criminal matters as long as one of the criminal acts or criminal consequences occurs in China.[45] In this provision, the phrase "in China" is interpreted to include Chinese ships and aircraft outside of the territory, as well as Chinese embassies and consulates which are located abroad.

Articles 7, 8 and 9 of the *Chinese Criminal Law*, respectively, stipulate personal jurisdiction, protection of jurisdiction and universal jurisdiction—all of which are considerations for determining the court of jurisdiction in criminal matters involving China.[46,47]

---

[42] *Criminal Law of the People's Republic of China* (adopted at the Second Session of the Fifth National People's Congress on 1 July 1979; revised at the Fifth Session of the Eighth National People's Congress on 14 March 1997 and promulgated by Order No.83 of the President of the People's Republic of China on 14 March 1997) https://www.ilo.org/dyn/natlex/docs/ELECTRONIC/5375/108071/F-78796243/CHN5375%20Eng3.pdf

[43] https://www.ilo.org/dyn/natlex/docs/ELECTRONIC/5375/108071/F-78796243/CHN5375%20Eng3.pdf

[44] *Criminal Law of the People's Republic of China* (adopted at the Second Session of the Fifth National People's Congress on 1 July 1979; revised at the Fifth Session of the Eighth National People's Congress on 14 March 1997 and promulgated by Order No.83 of the President of the People's Republic of China on 14 March 1997) https://www.ilo.org/dyn/natlex/docs/ELECTRONIC/5375/108071/F-78796243/CHN5375%20Eng3.pdf

[45] Xiaobing Li, Yongfeng Qin, "Research on Criminal Jurisdiction of Computer Cybercrime" in *Procedia Computer Science*, Volume 131, 2018, pp. 793–799. ISSN 1877–0509. https://doi.org/10.1016/j.procs.2018.04.263. (https://www.sciencedirect.com/science/article/pii/S1877050918306434).

[46] *Criminal Law of the People's Republic of China* (adopted at the Second Session of the Fifth National People's Congress on 1 July 1979; revised at the Fifth Session of the Eighth National People's Congress on 14 march 1997 and promulgated by Order No.83 of the President of the People's Republic of China on 14 March 1997) https://www.ilo.org/dyn/natlex/docs/ELECTRONIC/5375/108071/F-78796243/CHN5375%20Eng3.pdf

[47] https://www.ilo.org/dyn/natlex/docs/ELECTRONIC/5375/108071/F-78796243/CHN5375%20Eng3.pdf

The magnitude of the challenges associated with determining jurisdictional authority for criminal, and other activity, done using the Internet, becomes glaringly clear when we look at the massive global increase in cybercriminal activity and the reported rise of cyberattacks on government, infrastructure, corporations and individuals. On many occasions, these cyberattacks are found to have originated outside of the country in which the target is based, bringing the idea of international cyber law to the forefront of our global legal evolution.

3.5.2 Global Costs and Fatal Consequences

The costs of cybercrime include damage and destruction of data, stolen money, lost productivity, reputational harm, theft of intellectual property, theft of personal and financial data, embezzlement, fraud and identity theft, digital forensic investigation, restoration and deletion of hacked data and systems and post-attack disruption to the normal course of business.

In their *2021 Internet Crime Report*, the US Federal Bureau of Investigation (FBI) reported that the number of cybercrime complaints to the Federal Bureau of Investigation rose 7% in 2021 to 847,376 and total money lost to cybercrime increased 64% to $6.9 billion.[48,49]Among the 2021 complaints received, ransomware, business e-mail compromise schemes and the criminal use of cryptocurrency are among the top incidents reported.[50] In 2021 alone, business email compromise schemes resulted in 19,954 complaints with an adjusted loss of nearly $2.4 billion. In IBM's annual *Cost of a Data Breach Report*,[51] featuring research by the Ponemon Institute, IBM reported the highest average total data breach cost in its 17-year history, with data breach costs rising from USD 3.86 million to USD 4.24 million in 2021.[52]

In 2020, Cybersecurity Ventures, a leading cybersecurity research organisation, estimated the growth of global cybercrime to be about 15 percent per year, reaching a total cost of $10.5 trillion USD annually by 2025, up from $3 trillion USD in 201

[48] *Federal Bureau of Investigation Internet Crime Report 2021*. Available online at: https://www.ic3.gov/Media/PDF/AnnualReport/2021_IC3Report.pdf

[49] https://www.ic3.gov/Media/PDF/AnnualReport/2021_IC3Report.pdf

[50] *Federal Bureau of Investigation Internet Crime Report 2021*. Available online at: https://www.ic3.gov/Media/PDF/AnnualReport/2021_IC3Report.pdf

[51] https://www.ibm.com/security/data-breach

[52] IBM. *Cost of a Data Breach Report 2021*. https://www.ibm.com/security/data-breach.

5.[53,54] This would not only represent the greatest transfer of economic wealth in history[55] but is also exponentially larger than the damage inflicted from natural disasters in a year,[56] more profitable than the global trade of all major illegal drugs combined.[57]

Ransomware is one type of cyberattack that is especially known for being financially devastating for governments, organisations and corporations around the world. Adding to the financial impact, a September 2020 ransomware attack on the IT systems of a major Dusseldorf hospital led to the tragic death of a patient in life-threatening condition.[58] With IT systems crashing, the hospital was unable to access data or admit patients; surgeries were cancelled and emergency patients were taken to other hospitals for admission and treatment. For a patient in critical condition and in need of urgent medical treatment, who had to be rerouted to a hospital in a neighbouring city as a result of the ransomware attack, the hour-long delay in accessing treatment resulted in her death.

This is by far not the first time an emergency patient had to be rerouted to a different hospital as a result of a ransomware attack. According to data compiled by Emsisoft, 764 healthcare providers were victimised by ransomware in 2019, in the United States alone.[59]

---

[53] Morgan, Steve. "Cybercrime To Cost The World $10.5 Trillion Annually By 2025" in *Cyberwarfare In The C-Suite* a Special Report by *Cybercrime Magazine.* 13 November 2020. Available online at: https://cybersecurityventures.com/cybercrime-damages-6-trillion-by-2021/

[54] Pete Boden. "The Emerging Era of Cyber Defense and Cybercrime". *Microsoft Security Blog.* 27 January 2016. Available online at: https://www.microsoft.com/security/blog/2016/01/27/the-emerging-era-of-cyber-defense-and-cybercrime/

[55] "Protect your data through innovation" in AT&T Cybersecurity Insights: The CEO's Guide to Data Security, Volume 5. Available online at: https://1c7fab3im83f5gqiow2qqs2k-wpengine.netdna-ssl.com/wp-content/uploads/2021/01/attceocyberreport_compressed.pdf

[56] Rajinder Tumber. "Cyber Attacks: Igniting The Next Recession?" in *Forbes.* 5 January 2019. Available online at: https://www.forbes.com/sites/rajindertumber/2019/01/05/cyber-attacks-igniting-the-next-recession

[57] Kami Periman. "How to Prevent the Bank Robbery No One Can See" in *Cisco Blogs.* 29 August 2017. Available online at: https://blogs.cisco.com/financialservices/how-to-prevent-the-bank-robbery-no-one-can-see

[58] "German hospital hacked, patient taken to another city dies" in AP News. 17 September 2020. Available online at: https://apnews.com/article/technology-hacking-europe-cf8f8eee1adcec69bcc864f2c4308c94

[59] Emsisoft Malware Lab. "The State of Ransomware in the US: Report and Statistics 2019" in *Emsisoft Blogs.* 12 December 2019. Available online at: https://blog.emsisoft.com/en/34822/the-state-of-ransomware-in-the-us-report-and-statistics-2019/

3.5.3 Common Law Evolution of International Online Defamation

This 2022 case involved a family dispute resolution practitioner who alleged that she was defamed by an overseas man in several posts on the Internet. The court confirmed that the principles in *Dow Jones v Gutnick*[60] still apply in relation to a plaintiff proving that publication to some person other than the plaintiff has occurred and held that the plaintiff must plead and prove facts which establish that the material has been downloaded and viewed by someone.[61] The court also held that an inference of publication will generally not be drawn from the mere fact that material has been posted on the Internet, and a plaintiff must prove to whom material was published and which jurisdiction the material was downloaded in. Parts of this were also confirmed in the *Massarani v Kriz* decision, also in 2022.[62]

The *Newman v Whittington* case is particularly exciting as it was the first time that the NSW Supreme Court applied the novel "serious harm" test since its introduction in 2021.[63] The court held that a plaintiff is obliged to prove "serious harm" as a fact in every case. The question of exactly how serious the "serious harm" must be to satisfy the "serious harm test" remains unclear; however it seems that the court accepted the approach taken by the UK Supreme Court in the 2019 *Lachaux v Independent Print Ltd* case that serious harm is a question which must be determined by reference to actual facts of a publication's impact and not just to the meaning of the words.[64]

This case serves to illustrate the many complexities which are still faced by courts around the world, as the law and legal system continue to evolve in order to address the novel issues which arise in our online world.

International law structures relations among states and other international stakeholders through various prohibitions, requirements, and permissions. As such, it has provided a path for regulating global governance issues of all varieties. As nations continue to give increased attention to the governance of the internet and the technical architecture that allows the global internet to function, the role of

---

[60] *Dow Jones and Company Inc v Gutnick* [2002] HCA 56; 210 CLR 575; 194 ALR 433; 77 ALJR 255. https://adsdatabase.ohchr.org/IssueLibrary/HIGH%20COURT%20OF%20AUSTRALIA_Dow%20Jones%20and%20Company%20Inc%20v%20Gutnick.pdf

[61] *Dow Jones and Company Inc v Gutnick* [2002] HCA 56; 210 CLR 575; 194 ALR 433; 77 ALJR 255. https://adsdatabase.ohchr.org/IssueLibrary/HIGH%20COURT%20OF%20AUSTRALIA_Dow%20Jones%20and%20Company%20Inc%20v%20Gutnick.pdf

[62] *Massarani v Kriz* [2022] FCA 80. https://www.judgments.fedcourt.gov.au/judgments/Judgments/fca/single/2022/2022fca0080

[63] *Newman v Whittington* [2022] NSWSC 249. As per: https://www.natlawreview.com/article/and-so-it-begins-first-consideration-serious-harm-threshold-australian-defamation

[64] *Lachaux v Independent Print Ltd* [2019] UKSC 27, [2020] AC 612, [2019] 4 All ER 485, [2019] 3 WLR 18, 169 NLJ 7845, [2019] All ER (D) 42 (Jun). Available online at: https://www.lexiswebinars.co.uk/legal/tmt/defamation-claims-post-lachaux-2020/supporting-materials/Lachaux-v-Independent-Print-Limited-2019-UKSC-27.DOC.pdf

international law in the cyber context will continue to gain increasing prominence in our legislative discourse.

The two primary perspectives on this seem to be either for or against new law. While various states and stakeholders have suggested that existing international law is sufficient to regulate behaviour of states, others have suggested that there are gaps or inefficiencies in the existing law that require the formulation of new rules, via treaty or by an evolution of customary international law. Still within the second group, there exists a divide between those who would see a treaty created to protect states from people and those who would see the development of a treaty to protect people from states.

3.6 Summary

In this chapter, we discussed the concepts of jurisdictionality and how jurisdiction for a legal issue can be determined. We also discussed the personal connecting factors and determinants for jurisdiction to hear an issue. Finally, we explored international law and related legal strategies in dealing with conflicting laws between jurisdictions.

In the next chapter, we will explore the concept and realisation of digital governance and its intersection with data sovereignty on both a national and an international level.

The answers to the following questions are provided within this chapter:

1. What does it mean to be "domiciled" and why is it important in determining jurisdiction?
2. Why is it important to distinguish between jurisdiction, recognition and enforcement in international legal conflicts? What is the relationship between these three concepts?
3. What is the difference between personal jurisdiction, territorial jurisdiction and subject-matter jurisdiction?
4. How do original jurisdiction and appellate jurisdiction differ?
5. What is the difference between domestic law and international law?
6. What are some of the sources of international law?
7. What are some of the non-monetary global costs associated with cybercriminal activities?

Chapter 4
Technical Complexities

Ranging from increased legislation, larger data vaults, and the seemingly endless accumulation of data by corporations to issues of access to the Internet, digital rights, and net neutrality, the challenges arising in digital law are many and often overlap. The complexities of the technological world in which we have found ourselves are unprecedented in known human history. Never before have we been confronted with such vast quantities of data, accessible and available nearly on demand, and with such a significant economic drive. The relevance and importance of our data has never been greater than what we see today, as our world continues to shift online in terms of our work, our social lives, health care, and financial movements. This changing world has brought about new issues, both technical and ethical, in how we view our information and online activities with regard to sovereignty and governance in the digital world.

4.1 Topic Overview

In this chapter, we will discuss the current technical and legislative challenges which governments and corporations are facing around the world. We will discuss current challenges which are increasing the need for legislation around the world, such as those caused by increased data awareness and data vault, metadata, data ownership, and breaches. We will also examine the issues surrounding Internet governance including access and censorship, as well as sovereignty between nations and the global complexity of regulations within the online world.

M. Lukings, A. Habibi Lashkari, *Understanding Cybersecurity Law in Data Sovereignty and Digital Governance*, Progress in IS,
https://doi.org/10.1007/978-3-031-14264-2_4

4.2 Current Challenges

Cloud computing and cloud-service providers have brought about a significant shift in the globalised delivery of information technology services and computing resources. Most of us would likely be unsurprised to learn that the vast majority of our personal and professional data is stored in the cloud. The most popular social media apps—such as Instagram, Twitter, Facebook, and LinkedIn—all use cloud-based data storage for user accounts, profiles, uploaded content, and more. With the heavy volume of data being shared and the prevalence of these storage systems, we may at times overlook the principles we have been taught with regard to cyber hygiene and end up taking our data security for granted.

Around the world, public awareness of data sovereignty and related issues has increased in recent years, along with our collective ever-increasing reliance on cloud services and storage resources. As we rely more and more on our access to cloud data storage and other cloud-based services, the amount of storage that we collectively need will correspondingly increase. Not only does this endear us evermore to the third-party cloud-service providers on whom we rely so heavily, but it also presents us with an ever-increasing potential loss due to the increasing risk of massive data breaches, identity theft, digital exploitation, and widespread cyberattacks, among other cyber risks.

4.2.1 Increased Legislation

Increased data usage comes with increased legislation. What exactly this entails differs between nations and regions. Data is subject to the legislation of whatever country it resides in. This can introduce a lot of complex considerations into the mix. It may mean the laws to which an individual needs to adhere are different in relation to data than those laws to which that same individual needs to adhere within their specific country of residence. As we are most familiar with the laws which apply to us within our own country or region, the difference between the laws with respect to data can be confusing and unfamiliar.

It can also be problematic when we consider that data within the cloud is not actually static, but rather it is continuously moving between users spread across the world, often through the use of Application Programming Interfaces (or APIs). This means that the data will have to adhere to multiple regional legislations at the same time. Adding additional complexity to the mix is the issue that these national and international data laws are not always compatible with one another. In some cases, the laws may actually contradict. Resultantly, it can be very easy to get tangled up in a web of data protection infringements without even being aware of that risk in the first place. In some countries, this could end up with the involved parties facing a million-dollar civil lawsuit.

It is the penalties of not knowing the nuances of data regions that have begun to push businesses away from consolidated data storage providers and towards a collection of smaller and more localised storage providers. By combining several localised providers, a business can ensure that their approach to data management is refined enough to adequately comply with the various data legislation and individual laws which apply differently around the world.

As cloud computing has become key in business information sharing and storage, the focus has become on ensuring compliance with the laws and regulations wherever the data may be stored. When some company data can be accessed in foreign countries, those companies may face significant data sovereignty challenges and risk violating foreign laws while still being fully legally compliant within their own jurisdiction. For example, following the Snowden-WikiLeaks scandal in June 2013, many countries in the European Union ended up passing new regulations to establish a base level of data privacy and to better secure data transfers which move data outside of the European Union.

Within the banking and financial services industry, the issue of data sovereignty becomes more complex as these industries are obligated to adhere to strict compliance and regulation requirements as a result of the confidential and sensitive nature of the data held with respect to their business. As regulations in some countries are stricter than in others, it is necessary for banking and financial service companies to be fully aware of the location of their data and the levels of access to that data.

4.2.1.1 Examples

For example, India adopted the *Information Technology (Reasonable Security Practices and Procedures and Sensitive Personal Data or Information) Rules, 2011*, a strict privacy and security regulation which requires that any international information or sensitive data transfers "may be allowed only if it is necessary".[1] These rules also require that the same level of protection be provided in a foreign country whenever sensitive personal data may be transferred between those nations.

Another good example is the 2001 United States *Patriot Act*, which is concerned with the data privacy rights of United States citizens.[2] Among its many provisions, the *Patriot Act* gives the United States government the ability to legally access, intercept, and inspect "any tangible things (including books, records, papers, documents, and other items) for an investigation to protect against international terrorism or clandestine intelligence activities".

If a non-US firm, such as a Hong Kong company, uses a cloud application which runs outside of the United States but stores its data on US-based servers, that

[1] https://cis-india.org/internet-governance/files/it-reasonable-security-practices-and-procedures-and-sensitive-personal-data-or-information-rules-2011.pdf

[2] *USA PATRIOT Act: Preserving Life and Liberty* (Uniting and Strengthening America by Providing Appropriate Tools Required to Intercept and Obstruct Terrorism). 115 Stat. 272.

company may be caught under the United States *Patriot Act* regulations. In theory, this means that the United States government could have access to that data. This has caused issues between nations, as many Asian financial service firms no longer want to host their data in the United States and many common US-based financial tools cannot be used by Asian-based institutions without putting those institutions at risk of potential access to data by US entities. This results in a loss of business and increased inconvenience for banks and insurance companies and negatively impacts the growth of the cloud service industry, effectively preventing the creation of a collaborative global information security environment.

In recent years, countries in Asia have started to increase their enforcement of data security regulations. For example, the *Act on Real Name Financial Transactions and Guarantee of Secrecy* in South Korea[3] aims to protect data collected by financial services, and the *Act on the Protection of Personal Information* in Japan[4] imposes strong restrictions on the use and transfer of private data.

One solution for cloud service providers could be to set up a physical presence in multiple jurisdictions, enabling compliance with varying data security regulations between nations. As this could present issues of cost-effectiveness and scalability, another approach could be the "data-centric" approach, which would use permission policies to secure the data itself, regardless of the jurisdiction or platform, ensuring that the data could only be accessed within certain countries and restricting access entirely from other countries, thereby enabling continued security throughout the entire data cycle.

4.2.2 Increased Data Awareness

The greater emphasis on data within businesses and in combination with big milestones in the legal world—such as the implementation of the *General Data Protection Regulation* in the European Union—has heightened public attention and awareness of the issues inherent in the evolution of data privacy in the law. From this increase in awareness, we have seen an increase in customer expectations with regard to data privacy and accountability in how businesses manage their data. By working under a policy of transparency, companies can reassure customers of the security of their personal data.

Over the past few years, and particularly escalating during the COVID-19 pandemic, we have seen cyber data attacks reach an unprecedented level. This has catalysed a response from governments to ensure reporting of cyberattacks, data

[3] https://www.bnpparibas.co.kr/en/customer-due-diligence/#:~:text=The%20Real%20Name%20Financial%20Transactions,real%20name%20of%20the%20customers

[4] https://www.delphix.com/glossary/japan-act-protection-of-personal-information#:~:text=The%20Japan%20Act%20on%20the%20Protection%20of%20Personal%20Information%2C%20also,Information%20or%20risk%20legal%20action

breaches, and ransomware attacks, partially to increase the awareness of the need for greater security responsiveness and prevention of data breaches. As we continue to move our work, health, financial, and social information online, it has become increasingly necessary for everyone to understand the implications of having such data protected rather than readily accessible by outside parties.

Many data protection solutions focus on prevention and ignore a key aspect of risk management and compliance: attackers' access to sensitive data, which can reside on devices, applications, and/or in the cloud. Threat protection solutions, by contrast, identify attackers in the network but ignore the key aspect of security incidents: the sensitivity of data accessed during an attack. By unifying security and compliance under a new model of data-aware threat protection, businesses are better able to create trust while also reducing the associated risks to data and users.[5]

The data-aware threat protection model has four main advantages: risk-based incident prioritisation; more precise threat hunting; automatic remediation; and more effective security posture management.

1. **Risk-based incident prioritisation**
   Security operators typically prioritise incident response based on severity, but that neglects the overall business impact. Data classification awareness by threat protection solutions contributes to how alerts, incidents, and vulnerabilities are prioritised. It helps better determine the risk of the activity, which influences its prioritisation. An alert on a corporate device that stores sensitive data is more important than an alert on a device that does not store such data.
2. **More precise threat hunting**
   By tracing each attacker's action and intertwining it with the data classification context, analysts can better understand attackers' motivations and searches. This also arms hunters with the ability to reference data severity. Such context also enables better hunting for data exfiltration threats by understanding whether activity is malicious or benign. This is different in the case of sensitive data, however, as reading such a file may indicate anomalous access to sensitive data, copying a file may be part of staging for exfiltration, and screen capturing may be a way to steal sensitive data.
3. **Automatic remediation across security and compliance boundaries**
   Automation allows often understaffed security and compliance teams to do more and react more quickly, but missing the context of an incident overlooks vital information in remediation. Data classification awareness allows defenders to become more effective by defining custom response actions based on data sensitivity.
4. **More effective security posture management**
   Security and compliance teams should not just respond to data leaks or data exfiltration incidents after they occur; they should think about being proactive to reduce leaks. Knowing where sensitive data is stored and combining that with the

[5] https://www.darkreading.com/risk/data-awareness-is-key-to-data-security

compliance (data sensitivity) and security (risk) disciplines enables a proactive reduction of the chance and impact of data breaches.

4.2.3 Larger Data Vaults and Data Location

Businesses have never before held as much data as they do now. As data is reproduced so quickly, the volume of data continues to increase every day, necessitating greater availability and access to large-scale data storage vaults. The resulting outcome of handling increasingly larger volumes of data not only increases the required processing time but also presents an increase in the risk of human error and oversight.

In Canada, data sovereignty refers to the government of Canada's right to control access and disclosure permissions for its digital information, that is, data related to Canada—including Canadian businesses, organisations, and individuals—and subject only to Canadian laws. The government of Canada has begun to address the intersecting issues of cloud-based storage, data privacy, and data sovereignty by outlining a "cloud-first" policy strategy. This policy rests on top of the more standardised consumer protection and privacy laws which are already in place.

The "cloud-first" strategy was outlined in depth in the government of Canada's White Paper on *Data Sovereignty and Public Cloud*, which describes it as "a strategy whereby cloud services are identified and evaluated as the principal delivery option when initiating information technology investments, initiatives, strategies, and projects".[6] The "cloud-first" policy encourages government departments and related agencies to consider their use of cloud-based deployment models and to utilise data storage options in a prescribed order of priority, beginning with public cloud, followed by hybrid cloud, then private cloud, and finally non-cloud data storage models.

In comparison, Germany—a member nation of the European Union—has not established a specific legal framework for cloud-based services, meaning that those services are subject to the general laws of both Germany and the European Union, as Germany is a member nation of the European Union. The relevant laws which apply to data storage include the *German Civil Code*,[7] *Commercial Code*,[8] *Telemedia Act*,[9] and *Copyright Act*;[10] national information technology security laws; the European

[6] https://www.canada.ca/en/government/system/digital-government/digital-government-innovations/cloud-services/gc-white-paper-data-sovereignty-public-cloud.html

[7] https://www.gesetze-im-internet.de/englisch_bgb/

[8] https://www.investopedia.com/terms/h/hgb.asp#:~:text=Handelsgesetzbuch%20(HGB)%20is%20Germany's%20commercial,with%20the%20treatment%20of%20workers

[9] https://www.gesetze-im-internet.de/englisch_bsig/englisch_bsig.pdf

[10] https://www.gesetze-im-internet.de/englisch_urhg/englisch_urhg.html

Union's *General Data Protection Regulation* (GDPR);[11] and other standard rules relating to competition and consumer protection.

Contrasting with both Canada and Germany, the People's Republic of China does not permit cross-border data flows, as a general rule. This means that all companies operating in China, including all of the large-scale data handlers and critical information infrastructure providers, must specifically localise their data within China. For the very rare cases in which Chinese data might need to be transferred abroad, the transfer must be explicitly approved via a lengthy process through the designated approved authority: the Cyberspace Administration of China.

4.2.4 Metadata

Metadata is "data that provides information about other data", but not the content of the data, such as the text of a message or the image itself. It represents data about data. Metadata enriches the data with information that makes it easier to find, use, and manage. For instance, HTML tags define the layout for human readers. Semantic metadata helps computers to interpret data by adding references to concepts in a knowledge graph.

Metadata has various purposes. It can help users find relevant information and discover resources. It can also help organise electronic resources, provide digital identification, and archive and preserve resources. It allows users to access resources by "allowing resources to be found by relevant criteria, identifying resources, bringing similar resources together, distinguishing dissimilar resources, and giving location information.[12] "Metadata of telecommunication activities, including Internet traffic, is very widely collected by various national governmental organisations. This data is used for the purposes of traffic analysis and can be used for mass surveillance. [13]

Metadata was traditionally used in the card catalogues of libraries until the 1980s when libraries converted their catalogue data to digital databases. In the 2000s, as data and information were increasingly stored digitally, this digital data was described using metadata standards. There are many distinct types of metadata, including:

---

[11] *EU General Data Protection Regulation (GDPR):* Regulation (EU) 2016/679 of the European Parliament and of the Council of 27 April 2016 on the protection of natural persons with regard to the processing of personal data and on the free movement of such data, and repealing Directive 95/46/EC (General Data Protection Regulation), OJ 2016 L 119/1.

[12] National Information Standards Organization; Rebecca Guenther; Jaqueline Radebaugh (2004). *Understanding Metadata* (PDF). Bethesda, MD: NISO Press. ISBN 978–1–880,124-62-8.

[13] Bruce Schneier. "Metadata = Surveillance" in *Schneier on Security*, March/April 2014. Available online at: https://www.schneier.com/essays/archives/2014/03/metadata_surveillanc.html

Descriptive metadata—the descriptive information about a resource. It is used for discovery and identification.[14]

Structural metadata—metadata about containers of data and indicates how compound objects are put together, for example, how pages are ordered to form chapters. It describes the types, versions, relationships, and other characteristics of digital materials.[15]

Administrative metadata—the information to help manage a resource, like a resource type, permissions, and when and how it was created.[16,17]

Reference Metadata—The Information about the Contents and Quality of Statistical Data

Statistical metadata,[18] also called process data, may describe processes that collect, process, or produce statistical data.[19]

Legal metadata—provides information about the creator, copyright holder, and public licensing, if provided.

Metadata is not strictly bound to one of these categories, as it can describe a piece of data in many other ways. Metadata includes both data in motion (that which travels on the net) and data at rest (that which is stored on a hard drive or cloud). It gives part of the picture, not the full picture, when it comes to your data.

Now imagine what that metadata looks like when combined with other metadata. It can show how often you've emailed a particular person, whether you are sending large files, and the timing of your correspondence. Couple that even further with metadata related to where you have visited on the web or where the receiver has visited. A story will begin to form, and that story can be pretty telling, depending on your activity.

For data at rest, the implications are even greater. It is an easy target; it is stored and waiting to be viewed, read, and interpreted. There are three main types of metadata: descriptive, administrative, and structural.

- Descriptive metadata enables the discovery, identification, and selection of resources. It can include elements such as titles, authors, and subjects.
- Administrative metadata facilitates the management of resources. It can include elements such as technical, preservation, rights, and use.

[14]"Merriam Webster". Archived from the original on 27 February 2015.

[15]"An Architecture for Information in Digital Libraries". Archived from the original on 27 March 2017.

[16]Zeng, Marcia (2004). "Metadata Types and Functions". NISO. Archived from the original on 7 October 2016.

[17]National Information Standards Organization (NISO) (2001). *Understanding Metadata* (PDF). NISO Press. p. 1. ISBN 978–1–880,124-62-8.

[18]Dippo, Cathryn; Sundgren, Bo. "The Role of Metadata in Statistics" (PDF). *Bureau of Labor Statistics.*

[19]Directorate, OECD Statistics. "OECD Glossary of Statistical Terms - Reference metadata Definition". *stats.oecd.org.*

- Structural metadata, generally used in machine processing, describes relationships among various parts of a resource, such as chapters in a book.

Some examples of basic metadata are author, date created, date modified, and file size. Metadata is also used for unstructured data such as images, videos, web pages, spreadsheets, etc. Web pages often include metadata in the form of meta tags.

4.2.5 Data Ownership and Control

When we connect to the cloud, there are three major factors to consider: where the data is being stored, where the storage infrastructure is located, and which laws govern the access to and disclosure of the data stored on that cloud system. A cloud service provider—often a third party—is the entity that owns and operates the hardware, software, and infrastructure necessary for the cloud to be operational. These resources and the associated services can then be accessed over the Internet, on a pay-per-use basis, by individuals or organisations who subscribe to the service offered by the third party.

Cloud services are typically divided into three different models: public, private, and hybrid. In the public cloud model, organisations receive access to these services over the Internet from a shared pool of resources that are logically separated from one another. This is in contrast to the private cloud model, in which the cloud resources are dedicated solely to one organisation and are maintained on a private network. In the hybrid cloud model, both public and private clouds are used, with data and applications communicating between the two.

Cloud service subscribers usually have the option to isolate the storage of their data to a specific geographic region, allowing the subscriber to determine the nation under whose laws and policies the data is to be kept. The largest cloud service providers tend to deploy their services globally, giving the service provider the ability to move data anywhere in the world, as per the wishes of the subscriber. This is where we run into the challenge of data sovereignty.

When data is stored in a cloud environment, regardless of where the cloud-service resources are physically located, the data may be subject to the laws of other countries. In law, data sovereignty refers to the control a nation—or state—has over access and disclosure of its digital information, subject only to its laws. Where a cloud service provider—as a large-scale corporate entity—is operating on a global level, it could end up being required to comply with a court order, warrant, or subpoena request from a foreign law enforcement agency seeking to obtain data relating to the affairs of another nation.

If a national governing body decides to move its data outside of its territorial boundaries, that could have an impact on its ability to access data and services which are vital for the continuity of nationally based businesses, health care, and other important infrastructure. In this scenario, when it uses cloud-based storage located beyond its national borders, the governing body would not be able to ensure full sovereignty over its data. The negative outcomes which can stem from a lack of full data sovereignty could have the potential to create widespread damage to the government and the nation, as well as businesses, organisations, and individuals in that nation. In this case, sensitive or valuable data could also unknowingly be subject to foreign laws and/or be disclosed to a foreign government, potentially without any form of notice to those affected, whose data is being disclosed.

4.2.6 Data Breaches

As our computers and mobile devices get more connective features, there are more places for data to slip through. New technologies are being created faster than we can protect them.

A **data breach** is an incident where information is stolen or taken from a system without the knowledge or authorisation of the system's owner. It exposes confidential, sensitive, or protected information to an unauthorised person. The files in a data breach are viewed and/or shared without permission. In general, data breaches happen due to weaknesses in technology or user behaviour, and most of them fall under five categories, including hacks, employee error, accidental exposure, physical theft, and third-party subcontractors.

Anyone can be at risk of a data breach—from individuals to high-level enterprises and governments. More importantly, anyone can put others at risk if they are not protected. The effects brought on by a data breach can come in the form of damage to the target company's reputation due to a perceived "betrayal of trust". Victims and their customers may also suffer financial losses should related records be part of the information stolen.

For example, in an SEC filing made on April 4, 2022,[20] Block—the company formerly known as Square—acknowledged that "Cash App" had been breached by a former employee in December of 2021. The leak included customers' names, brokerage account numbers, and other data, such as portfolio value and stock trading activity. Block has not been forthcoming about how many customers were affected in total, but the company is contacting over eight million customers to inform them about the incident. Based on what they have said so far, no other personally identifiable information or account credentials were leaked in the incident.

[20] Firewall Times, The Most Recent Data Breaches – April 2022, https://firewalltimes.com/recent-data-breaches/

4.2.7 Cyber Exceptionalism

Cyber exceptionalism suggests that the digital realm is qualitatively distinctive from the analogue world and that digital spaces, therefore, need to be treated differently from all previous technological innovations. This perspective was especially popular during the rise of the commercial Internet, in the 1990s, but is still evident in public and academic discourse today. In cyberspace, cyber exceptionalism suggests a similar distinctiveness of culture, approach, and even tactics and strategy. Cyber exceptionalism is the process by which the application of the prefix "cyber" to any concept x, such as space, thereby opens up an analytical difference between the meaning of cyber-x and x.

Cyber exceptionalist thinking is based on the assumption that the growing importance of computer-aided network communication implies the demise of state sovereignty.[21] Although the Internet's actual development did not take place outside of concrete legal spaces and would not have been possible without the incentives provided by markets, regulatory regimes, or public research infrastructures,[22] cyber exceptionalism—which most often takes the form of cyber libertarianism[23]—was the formative ideology in those early days, with a strong cultural and economic backing in Silicon Valley.[24]

The idea of state sovereignty was particularly challenged by two different—yet related—discursive strands, namely cyber exceptionalism and multi-stakeholder Internet governance, that significantly shaped public and academic discourses. Yet, in more recent years, policy actors have successfully sought to justify and reaffirm sovereignty in the digital sphere against these two perspectives.

The idea of cyber exceptionalism is important for those advocating that "cyber" questions require an entirely separate approach. Just as the discipline of international relations came into being because the international system was perceived to be fundamentally different from domestic society, at the core of cyber exceptionalism is the belief that the growth of cyberspace creates distinct dynamics worthy of placing into a new field of study.

---

[21] Katz, J. (1997). Birth of a Digital Nation. In Wired. https://www.wired.com/1997/04/netizen-3/

[22] Mazzucato, M. (2011). The entrepreneurial state. Demos. http://oro.open.ac.uk/30159/1/Entrepreneurial_State_-_web.pdf

[23] Keller, C. I. (2019). Exception and Harmonization: Three Theoretical Debates on Internet Regulation (2020(2); HIIG Discussion Paper Series). Alexander von Humboldt Institut für Internet und Gesellschaft. https://doi.org/10.2139/ssrn.3572763

[24] Barbrook, R., & Cameron, A. (1996). The Californian ideology. Science as Culture, 6(1), 44–72. https://doi.org/10.1080/09505439609526455

4.3 Internet Governance

Internet governance refers to the development and application of shared principles, norms, rules, decision-making procedures, and programs that work in combination to shape the evolution and use of the Internet.

A working group established after a UN-initiated World Summit on the Information Society (WSIS) proposed the following definition of Internet governance as part of its June 2005 report:

> Internet governance is the development and application by governments, the private sector, and civil society, in their respective roles, of shared principles, norms, rules, decision-making procedures, and programmes that shape the evolution and use of the Internet.

Law professor Yochai Benkler developed a conceptualisation of Internet governance by the idea of three "layers" of governance:

1. Physical infrastructure layer (through which information travels).
2. Code or logical layer (controls the infrastructure).
3. Content layer (contains the information signalled through the network).

Professors Jovan Kurbalija and Laura DeNardis also offer comprehensive definitions of "Internet Governance". According to Kurbalija, the broad approach to Internet governance goes "beyond Internet infrastructural aspects and addresses other legal, economic, developmental, and sociocultural issues"; along similar lines, DeNardis argues that "Internet Governance generally refers to the policy and technical coordination issues related to the exchange of information over the Internet". One of the more policy-relevant questions today is exactly whether the regulatory responses are appropriate to police the content delivered through the Internet: they include important rules for the improvement of Internet safety and for dealing with threats such as cyber-bullying, copyright infringement, data protection, and other illegal or disruptive activities.

There is no one single person, company, organisation, or government that runs the Internet. Rather, the Internet is a massive globally distributed network composed of many voluntarily interconnected autonomous networks. The Internet operates without a central governing body, with each constituent network setting and enforcing its own policies. Internet governance is conducted and overseen by a decentralised and international multi-stakeholder network of interconnected autonomous groups drawing from civil society, the private sector, governments, the academic and research communities, and national and international organisations. These autonomous groups work cooperatively, from their respective roles, to create shared policies and standards that maintain the global interoperability of the Internet for the public good.

The definition of Internet governance has been contested by differing groups across political and ideological lines. One of the main debates concerns the authority and participation of certain actors, such as national governments, corporate entities, and civil society, to play a role in the Internet's governance.

For example, in an effort to help ensure interoperability, several key technical and policy aspects of the underlying core infrastructure and the principal namespaces are administered by the Internet Corporation for Assigned Names and Numbers (ICANN). Headquartered in Los Angeles, California, ICANN is governed by an international board of directors drawn from across the Internet's technical, business, academic, and other non-commercial communities. Seeking to create a globally unified namespace to ensure and protect the global reach of the Internet, ICANN is responsible for overseeing the assignment of globally unique identifiers on the Internet, including domain names, Internet protocol addresses, application port numbers in the transport protocols, and many other parameters.

The technical underpinning and standardisation of core protocols of the Internet is an activity of the Internet Engineering Task Force (IETF), a non-profit organisation of loosely affiliated international participants that anyone may associate with by contributing technical expertise.

4.3.1 Access

Availability of Internet access was once limited but has grown rapidly. In 1995, only 0.04 percent of the world's population had access, with well over half of those living in the United States, and consumer use was primarily through dial-up. By the first decade of the twenty-first century, many consumers in developed nations used faster broadband technology, and by 2014, 41 percent of the world's population had access, broadband was almost ubiquitous worldwide, and global average connection speeds exceeded one megabit per second.

Internet access refers to the ability of individuals and organisations to connect to the Internet using computer terminals, computers, and other devices; and to access services such as email and the World Wide Web. Internet access is sold by Internet service providers (ISPs) delivering connectivity at a wide range of data transfer rates via various networking technologies. Many organisations, including a growing number of municipal entities, also provide cost-free wireless Internet access as well as landline access.

4.3.1.1 Examples

Several countries have adopted laws that require the state to work to ensure that Internet access is broadly available and/or prevent the state from unreasonably restricting an individual's access to information and the Internet:

- **Costa Rica**

  A 30 July 2010 ruling by the Supreme Court of Costa Rica stated: "Without fear of equivocation, it can be said that these technologies [information technology and communication] have impacted the way humans communicate, facilitating the connection between people and institutions worldwide and eliminating barriers of space and time. At this time, access to these technologies has become a basic tool to facilitate the exercise of fundamental rights and democratic participation (e-democracy) and citizen control, education, freedom of thought and expression, access to information and public services online, the right to communicate with the government electronically and administrative transparency, among others. This includes the fundamental right of access to these technologies, in particular, the right of access to the Internet or World Wide Web".

- **Estonia**
  In 2000, the parliament launched a massive program to expand access to the countryside. The Internet, the government argues, is essential for life in the twenty-first century.
- **Finland**
  By July 2010, every person in Finland was to have access to a one-megabit per second broadband connection, according to the Ministry of Transport and Communications. And by 2015, access to a 100 Mbit/s connection.
- **France**
  In June 2009, the Constitutional Council, France's highest court, declared access to the Internet to be a basic human right in a strongly worded decision that struck down portions of the HADOPI law, a law that would have tracked abusers without judicial review and automatically cut off network access to those who continued to download illicit material after two warnings.
- **Greece**
  Article 5A of the Constitution of Greece states that all persons have the right to participate in the Information Society and that the state has an obligation to facilitate the production, exchange, diffusion, and access of electronically transmitted information.
- **Spain**
  Starting in 2011, Telefónica, the former state monopoly that holds the country's "universal service" contract, has to guarantee to offer "reasonably" priced broadband of at least one megabyte per second throughout Spain.

4.3.2 Censorship

Internet censorship refers to the control or suppression of what can be accessed, published, or viewed on the Internet enacted by regulators, or on their own initiative. Internet censorship puts restrictions on what information can be put on the Internet or not. Individuals and organisations may engage in self-censorship for moral, religious, or business reasons, to conform to societal norms, due to intimidation, or out of fear of legal or other consequences.

The extent of Internet censorship varies on a country-to-country basis. While some countries have moderate Internet censorship, other countries go as far as to limit access to information such as news and suppress and silence discussion among citizens. Internet censorship also occurs in response to or in anticipation of events such as elections, protests, and riots. An example is the increased censorship due to the events of the Arab Spring. Other types of censorship include the use of copyrights, defamation, harassment, and various obscene material claims as a way to deliberately suppress content.

Many of the challenges associated with Internet censorship are similar to those for offline censorship of more traditional media such as newspapers, magazines, books, music, radio, television, and film. One difference is that national borders are more permeable online: residents of a country that bans certain information can find it on websites hosted outside the country. Thus censors must work to prevent access to information even though they lack physical or legal control over the websites themselves. This in turn requires the use of technical censorship methods that are unique to the Internet, such as site blocking and content filtering.

Support for and opposition to Internet censorship also varies. In a 2012 Internet Society survey, 71% of respondents agreed that "censorship should exist in some form on the Internet". In the same 2012 survey, 83% of respondents agreed that "access to the Internet should be considered a basic human right" and 86% of respondents agreed that "freedom of expression should be guaranteed on the Internet". Perception of Internet censorship in the United States is largely based on the First Amendment and the right to expansive free speech and access to content without regard to the consequences. According to GlobalWebIndex, over 400 million people use virtual private networks to circumvent censorship or for increased user privacy.

Views about the feasibility and effectiveness of Internet censorship have evolved in parallel with the development of the Internet and censorship technologies:

- A 1993 Time Magazine article quotes computer scientist John Gilmore, one of the founders of the Electronic Frontier Foundation, as saying, "The Net interprets censorship as damage and routes around it".
- In November 2007, "Father of the Internet" Vint Cerf stated that he sees government control of the Internet failing because the Web is almost entirely privately owned.
- A report of research conducted in 2007 and published in 2009 by the Berkman Center for Internet & Society at Harvard University stated that: "We are confident that the [censorship circumvention] tool developers will, for the most part, keep ahead of the governments' blocking efforts", but also that "...we believe that less than two percent of all filtered Internet users use circumvention tools".
- In contrast, a 2011 report by researchers at the Oxford Internet Institute published by UNESCO concludes "...the control of information on the Internet and Web is certainly feasible, and technological advances do not therefore guarantee greater freedom of speech".

Blocking and filtering can be based on relatively static blacklists or be determined more dynamically based on a real-time examination of the information being exchanged. Blacklists may be produced manually or automatically and are often not available to non-customers of the blocking software. Blocking or filtering can be done at a centralised national level, at a decentralised sub-national level, or at an institutional level, for example in libraries, universities, or Internet cafes. Blocking and filtering may also vary within a country across different ISPs. Countries may filter sensitive content on an ongoing basis and/or introduce temporary filtering during key time periods such as elections. In some cases, the censoring authorities may surreptitiously block content to mislead the public into believing that censorship has not been applied. This is achieved by returning a fake "Not Found" error message when an attempt is made to access a blocked website.

Unless the censor has total control over all Internet-connected computers, such as in North Korea (which employs an intranet that only privileged citizens can access), or Cuba, total censorship of information is very difficult or impossible to achieve due to the underlying distributed technology of the Internet. Pseudonymity and data havens (such as Freenet) protect free speech using technologies that guarantee material cannot be removed and prevent the identification of authors. Technologically savvy users can often find ways to access blocked content. Nevertheless, blocking remains an effective means of limiting access to sensitive information for most users when censors, such as those in China, are able to devote significant resources to building and maintaining a comprehensive censorship system.

4.3.2.1 Technical Censorship

Various parties are using different technical methods of preventing public access to undesirable resources, with varying levels of effectiveness, costs, and side effects. These include:

Method One: Blacklist

Entities mandating and implementing censorship usually identify them by one of the following items: keywords, domain names, or IP addresses. Lists are populated from different sources, ranging from private suppliers through courts to specialised government agencies.

As per Hoffmann, different methods are used to block certain websites or pages including DNS poisoning, blocking access to IPs, analysing and filtering URLs, inspecting filter packets, and resetting connections.

Method Two: Points of Control

Enforcement of the censor-nominated technologies can be applied at various levels of countries and Internet infrastructure:

- **Internet backbone**—including Internet exchange points (IXP) with international networks, operators of submarine communications cables, satellite Internet access points, international optical fibre links, etc. In addition to facing huge performance challenges due to the large bandwidths involved, these do not give censors access to information exchanged within the country.
- **Internet service providers (ISPs)**—which involves the installation of voluntary (as in the United Kingdom) or mandatory (as in Russia) Internet surveillance and blocking equipment.
- **Individual institutions**—which in most cases implement some form of Internet access controls to enforce their own policies, but, especially in the case of public or educational institutions, may be requested or coerced to do this on the request of the government.
- **Personal devices**—whose manufacturers or vendors may be required by law to install censorship software.
- **Application service providers** (e.g. social media companies)—who may be legally required to remove particular content. Foreign providers with a business presence in a given country may be also coerced into restricting access to specific contents for visitors from the requesting country.
- **Certificate authorities**—may be required to issue counterfeit X.509 certificates controlled by the government, allowing man-in-the-middle surveillance of TLS encrypted connections.
- **Content Delivery Network providers**—who tend to aggregate large amounts of content (e.g. images) may also be attractive targets for censorship authorities.

Method Three: Internet Protocol Address Blocking

Access to a certain IP address is denied. If the target website is hosted on a shared hosting server, all websites on the same server will be blocked. This affects IP-based protocols such as HTTP, *F*TP, and POP. A typical circumvention method is to find proxies that have access to the target websites, but proxies may be jammed or blocked, and some websites, such as Wikipedia (when editing), also block proxies. Some large websites such as Google have allocated additional IP addresses to circumvent the block, but later the block was extended to cover the new addresses. Due to challenges with geolocation, geo-blocking is normally implemented via IP address blocking.

Method Four: Domain Name System Filtering and Redirection

Blocked domain names are not resolved, or an incorrect IP address is returned via DNS hijacking or other means. This affects all IP-based protocols such as HTTP, FTP. and POP. A typical circumvention method is to find an alternative DNS resolver that resolves domain names correctly, but domain name servers are subject to blockage as well, especially IP address blocking. Another workaround is to bypass DNS if the IP address is obtainable from other sources and is not itself blocked. Examples are modifying the Hosts file or typing the IP address instead of the domain name as part of a URL given to a Web browser.

Method Five: Uniform Resource Locator (URL) Filtering

URL strings are scanned for target keywords regardless of the domain name specified in the URL. This affects the HTTP protocol. Typical circumvention methods are to use escaped characters in the URL or to use encrypted protocols such as VPN and TLS/SSL.

Method Six: Packet Filtering

Terminate TCP packet transmissions when a certain number of controversial keywords are detected. This affects all TCP-based protocols such as HTTP, FTP. and POP, but search engine results pages are more likely to be censored. Typical circumvention methods are to use encrypted connections—such as VPN and TLS/SSL—to escape the HTML content, or by reducing the TCP/IP stack's MTU/MSS to reduce the amount of text contained in a given packet.

Method Seven: Connection Reset

If a previous TCP connection is blocked by the filter, future connection attempts from both sides can also be blocked for some variable amount of time. Depending on the location of the block, other users or websites may also be blocked, if the communication is routed through the blocking location. A circumvention method is to ignore the reset packet sent by the firewall.

Method Eight: Network Disconnection

A technically simpler method of Internet censorship is to completely cut off all routers, either by software or by hardware (turning off machines, pulling out cables). A circumvention method could be to use a satellite ISP to access the Internet.

Method Nine: Portal Censorship and Search Result Removal

Major portals, including search engines, may exclude websites that they would ordinarily include. This renders a site invisible to people who do not know where to find it. When a major portal does this, it has a similar effect to censorship. Sometimes this exclusion is done to satisfy a legal or other requirement; other times, it is purely at the discretion of the portal. For example, Google.de and Google.fr remove Neo-Nazi and other listings in compliance with German and French law.

Method Ten: Computer Network Attacks

Denial-of-service attacks and attacks that deface opposition websites can produce the same result as other blocking techniques, preventing or limiting access to certain websites or other online services, although only for a limited period of time. This technique might be used during the lead up to an election or some other sensitive period. It is more frequently used by non-state actors seeking to disrupt services.

4.3.2.2 Non-Technical Censorship

Internet content is also subject to censorship methods similar to those used with more traditional media. For example:

- Laws and regulations may prohibit various types of content and/or require that content be removed or blocked either proactively or in response to requests.
- Publishers, authors, and ISPs may receive formal and informal requests to remove, alter, slant, or block access to specific sites or content.
- Publishers and authors may accept bribes to include, withdraw, or slant the information they present.
- Publishers, authors, and ISPs may be subject to arrest, criminal prosecution, fines, and imprisonment.
- Publishers, authors, and ISPs may be subject to civil lawsuits.
- Equipment may be confiscated and/or destroyed.
- Publishers and ISPs may be closed or required licences may be withheld or revoked.
- Publishers, authors, and ISPs may be subject to boycotts.
- Publishers, authors, and their families may be subject to threats, attacks, beatings, and even murder.
- Publishers, authors, and their families may be threatened with or actually lose their jobs.
- Individuals may be paid to write articles and comments in support of particular positions or to attack opposition positions, usually without acknowledging the payments to readers and viewers.
- Censors may create their own online publications and websites to guide online opinion.
- Access to the Internet may be limited due to restrictive licensing policies or high costs.
- Access to the Internet may be limited due to a lack of the necessary infrastructure, deliberate or not.

- Access to search results may be restricted due to government involvement in the censorship of specific search terms. Content may be excluded due to terms set with search engines. By allowing search engines to operate in new territory, they must agree to abide by censorship standards set by the government of that country.

4.3.2.3 Censorship of Users

In the effort to censor users in online forums, there are a variety of tools and options available. Two common methods used by online corporations to encourage censorship among platform users are through deplatforming and more traditional terms of service agreements. When a user violates the terms of service, they can be deplatformed by the online forum itself, as a method of censorship and control.

Method One: Deplatforming

Deplatforming is a form of Internet censorship in which controversial speakers or speech are suspended, banned, or otherwise shut down by social media platforms and other service providers that generally provide a venue for free speech or expression. Banking and financial service providers, among other companies, have also denied services to controversial activists or organisations, a practice known as "financial deplatforming".

Law professor Glenn Reynolds dubbed 2018 the "Year of Deplatforming", in an August 2018 article in The Wall Street Journal. According to Reynolds, in 2018, "the Internet giants decided to slam the gates on a number of people and ideas they don't like". On 6 August 2018, for example, several major platforms, including YouTube and Facebook, executed a coordinated, permanent ban on all accounts and media associated with conservative talk show host Alex Jones and his media platform InfoWars, citing "hate speech" and "glorifying violence".

Method Two: Terms of Service Agreements

Most major web service operators reserve to themselves broad rights to remove or pre-screen content, and to suspend or terminate user accounts, sometimes without giving a specific list or only a vague general list of the reasons allowing the removal. The phrases "at our sole discretion", "without prior notice", and "for other reasons" are common in Terms of Service agreements.

- **Facebook**
  Among other things, the Facebook Statement of Rights and Responsibilities says: "You will not post content that: is hateful, threatening, or pornographic; incites violence; or contains nudity or graphic or gratuitous

violence", "You will not use Facebook to do anything unlawful, misleading, malicious, or discriminatory", "We can remove any content or information you post on Facebook if we believe that it violates this Statement", and "If you are located in a country embargoed by the United States, or are on the U.S. Treasury Department's list of Specially Designated Nationals, you will not engage in commercial activities on Facebook (such as advertising or payments) or operate a Platform application or website".

- **Google**
  Google's general Terms of Service, which were updated on 1 March 2012, state: "We may suspend or stop providing our Services to you if you do not comply with our terms or policies or if we are investigating suspected misconduct", "We may review content to determine whether it is illegal or violates our policies, and we may remove or refuse to display content that we reasonably believe violates our policies or the law", and "We respond to notices of alleged copyright infringement and terminate accounts of repeat infringers according to the process set out in the U.S. Digital Millennium Copyright Act".
- **Google Search**
  Google's Webmaster Tools help includes the following statement: "Google may temporarily or permanently remove sites from its index and search results if it believes it is obligated to do so by law, if the sites do not meet Google's quality guidelines, or for other reasons, such as if the sites detract from users' ability to locate relevant information".
- **Twitter**
  The Twitter Terms of Service state: "We reserve the right at all times (but will not have an obligation) to remove or refuse to distribute any Content on the Services and to terminate users or reclaim usernames" and "We reserve the right to remove Content alleged to be copyright infringing without prior notice and at our sole discretion".
- **YouTube**
  The YouTube Terms of Service include the following statements: "YouTube reserves the right to decide whether Content violates these Terms of Service for reasons other than copyright infringement, such as, but not limited to, pornography, obscenity, or excessive length. YouTube may at any time, without prior notice and in its sole discretion, remove such Content and/or terminate a user's account for submitting such material in violation of these Terms of Service", "YouTube will remove all Content if properly notified that such Content infringes on another's intellectual property rights", and "YouTube reserves the right to remove Content without prior notice".
- **Wikipedia**
  Content within a Wikipedia article may be modified or deleted by any editor as part of the normal process of editing and updating articles. All editing decisions are open to discussion and review. The Wikipedia Deletion policy outlines the circumstances in which entire articles can be deleted. Any editor who believes a page doesn't belong in an encyclopaedia can propose its deletion. Such a page can be deleted by any administrator if, after seven days, no one objects to the

proposed deletion. Speedy deletion allows for the deletion of articles without discussion and is used to remove pages that are so obviously inappropriate for Wikipedia that they have no chance of surviving a deletion discussion. All deletion decisions may be reviewed, either informally or formally.

- **Yahoo!**

  Yahoo!'s Terms of Service (TOS) state: "You acknowledge that Yahoo! may or may not pre-screen Content, but that Yahoo! and its designees shall have the right (but not the obligation) in their sole discretion to pre-screen, refuse, or remove any Content that is available via the Yahoo! Services. Without limiting the foregoing, Yahoo! and its designees shall have the right to remove any Content that violates the TOS or is otherwise objectionable".

4.3.2.4 Circumvention

Internet censorship circumvention is the processes used by technologically savvy Internet users to bypass the technical aspects of Internet filtering and gain access to otherwise censored material. Circumvention is an inherent problem for those wishing to censor the Internet because filtering and blocking do not remove content from the Internet, but instead block access to it. Therefore, as long as there is at least one publicly accessible uncensored system, it will often be possible to gain access to the otherwise censored material. However, circumvention may not be possible for non-tech-savvy users, so blocking and filtering remain effective means of censoring the Internet access of large numbers of users.

Different techniques and resources are used to bypass Internet censorship, including proxy websites, Virtual Private Networks (VPN),[25] "sneakernets", the dark web,[26] and circumvention software tools. Solutions have differing ease of use, speed, security, and risks. Most, however, rely on gaining access to an Internet connection that is not subject to filtering, often in a different jurisdiction not subject to the same censorship laws. According to GlobalWebIndex????, over 400 million people use virtual private networks to circumvent censorship or for an increased level of privacy. The majority of circumvention techniques are not suitable for day-to-day use.

There are risks to using circumvention software or other methods to bypass Internet censorship. In some countries, individuals that gain access to otherwise

[25] "Characterization of Encrypted and VPN Traffic Using Time-Related Features", Gerard Drapper Gil, Arash Habibi Lashkari, Mohammad Mamun, Ali A. Ghorbani, In Proceedings of the second International Conference on Information Systems Security and Privacy(ICISSP 2016), pages 407–414, Rome, Italy.

[26] "DIDarknet: A Contemporary Approach to Detect and Characterize the Darknet Traffic using Deep Image Learning", Arash Habibi Lashkari, Gurdip Kaur, Abir Rahal, The tenth International Conference on Communication and Network Security (ICCNS2020), Tokyo, Japan, September 2020.

restricted content may be violating the law and if caught can be expelled, fired, jailed, or subject to other punishments and loss of access.

In June 2011, the New York Times[27] reported that the United States is engaged in a "global effort to deploy 'shadow' Internet and mobile phone systems that dissidents can use to undermine repressive governments that seek to silence them by censoring or shutting down telecommunications networks".[28]

Another way to circumvent Internet censorship is to physically go to an area where the Internet is not censored. In 2017, a so-called "Internet refugee camp" was established by IT workers in the village of Bonako, just outside an area of Cameroon where the Internet is regularly blocked.[29]

4.3.2.5 Primarily Targeted Groups/Themes

Protecting existing economic interests is an additional emergent motive for Internet filtering. In addition, networking tools and applications that allow the sharing of information related to these motives are themselves subjected to filtering and blocking. And while there is considerable variation from country to country, the blocking of websites in a local language is roughly twice that of websites available only in English or other international languages. There are several motives or rationales for Internet filtering. Some of the major driving forces that we will discuss include politics and power; social norms and morals; security concerns; existing economic interests and copyright; network tools; and individual personal information.

Politics and Power

Censorship directed at political opposition to the ruling government is common in authoritarian and repressive regimes. Some countries block websites related to religion and minority groups, often when these movements represent a threat to the ruling regimes. Examples include:

- Political blogs and websites
- *Lèse majesté* sites, sites with content that offends the dignity of or challenges the authority of a reigning sovereign or of a state
- Falun Gong and Tibetan exile group sites in China or Buddhist, Cao Dai faith, and indigenous hill tribe sites in Vietnam

[27] https://www.nytimes.com/2011/06/12/world/12internet.html

[28] https://www.nytimes.com/2011/06/12/world/12internet.html

[29] Dahir, Abdi Latif. "Reeling from an internet shutdown, startups in Cameroon have created an "internet refugee camp"". *Quartz Africa*. Available at: https://qz.com/africa/942879/an-internet-shutdown-in-cameroon-has-forced-startups-to-create-an-internet-refugee-camp-in-bonako-village/

- 50 Cent Party, or "50 Cent Army" that worked to sway negative public opinion on the Communist Party of China
- Russian web brigades
- Sites aimed at religious conversion from Islam to Christianity
- Sites criticising the government or any authority in the country
- Sites that comment on political parties that oppose the current government of a country
- Sites that accuse authorities of corruption
- Sites that comment on minorities or LGBT issues

Social Norms and Morals

Social filtering is the censorship of topics that are held to be antithetical to accepted societal norms. In particular, censorship of child pornography and the desire to protect children enjoys very widespread public support, and such content is subject to censorship and other restrictions in most countries. Examples include:

- Sites that include hate speech inciting racism, sexism, homophobia, or other forms of bigotry
- Sites seen as promoting illegal drug use
- Sex and erotic, fetishism, prostitution, and pornographic sites
- Child pornography and pedophile-related sites
- Gambling sites
- Sites encouraging or inciting violence
- Sites promoting criminal activity
- Communist symbols and imagery in Poland, Lithuania, Ukraine, Latvia, Moldova, and Hungary
- Nazi and similar websites, particularly in France and Germany
- Sites that contain blasphemous content, particularly when directed at a majority or state-supported religion
- Sites that contain defamatory, slanderous, or libellous content
- Sites that include political satire
- Sites that contain information on social issues or "online protests, petitions, and campaigns"

Security Concerns

Many organisations implement filtering as part of a defence in depth strategy to protect their environments from malware and to protect their reputations in the event of their networks being used, for example, to carry out sexual harassment.

Internet filtering related to threats to national security that targets the websites of insurgents, extremists, and terrorists often enjoys wide public support. Examples include:

- Blocking of pro–North Korean sites by South Korea.
- Blocking sites of groups that foment domestic conflict in India.
- Blocking of sites of the Muslim Brotherhood in some countries in the Middle East.
- Blocking WikiLeaks.
- Blocking sites such as 4chan thought to be related to the group Anonymous.

Existing Economic Interests and Copyrights

The protection of existing economic interests is sometimes the motivation for blocking new Internet services such as low-cost telephone services that use Voice over Internet Protocol (VoIP). These services can reduce the customer base of telecommunications companies, many of which enjoy entrenched monopoly positions and some of which are government sponsored or controlled.

Anti-copyright activists Christian Engström, Rick Falkvinge, and Oscar Swartz have alleged that censorship of child pornography is being used as a pretext by copyright lobby organisations to get politicians to implement similar site-blocking legislation against copyright-related piracy. Examples include:

- File sharing and peer-to-peer (P2P) related websites such as The Pirate Bay.
- Skype.
- Sites that sell or distribute music, but are not approved by rights holders, such as allofmp3.

Network Tools

Blocking the intermediate tools and applications of the Internet that can be used to assist users in accessing and sharing sensitive material is common in many countries. Examples include:

- Media sharing websites (e.g. Flickr and YouTube).
- Social networks (e.g. Facebook and Instagram).
- Translation sites and tools.
- Email providers.
- Web hosting sites.
- Blog hosting sites such as Blogspot and MediuM.
- Microblogging sites such as Twitter and Weibo.
- Wikipedia.
- Censorship circumvention sites, such as anonymizers and proxy avoidance sites.
- Search engines such as Bing and Google—particularly in Mainland China and Cuba.

Individual Personal Information

The right to be forgotten is a concept that has been discussed and put into practice in the European Union. In May 2014, the European Court of Justice ruled against Google in *Costeja*, a case brought by a Spanish man who requested the removal of a link to a 1998 article in *La Vanguardia* newspaper that had been digitised, about an auction for his foreclosed home, for a debt that he had subsequently paid.[30] He initially attempted to have the article removed by complaining to Spain's data protection agency—Agencia Española de Protección de Datos—which rejected the claim on the grounds that it was lawful and accurate, but accepted a complaint against Google and asked Google to remove the results. Google sued in Spain, and the lawsuit was transferred to the European Court of Justice. The court ruled in *Costeja* that search engines are responsible for the content they point to, and thus, Google was required to comply with EU data privacy laws. It began compliance on May 30, 2014, during which it received 12,000 requests to have personal details removed from its search engine.

Index on Censorship claimed that "*Costeja* ruling. . .allows individuals to complain to search engines about information they do not like with no legal oversight. This is akin to marching into a library and forcing it to pulp books. Although the ruling is intended for private individuals it opens the door to anyone who wants to whitewash their personal history. The Court's decision is a retrograde move that misunderstands the role and responsibility of search engines and the wider Internet. It should send chills down the spine of everyone in the European Union who believes in the crucial importance of free expression and freedom of information".[31]

4.3.2.6 Resilience

Various contexts influence whether an Internet user will be resilient to censorship attempts. Users are more resilient to censorship if they are aware that information is being manipulated. This awareness of censorship leads to users finding ways to circumvent it. Awareness of censorship also allows users to factor this manipulation into their belief systems. Knowledge of censorship also offers some citizens an incentive to try to discover information that is being concealed. In contrast, those that lack awareness of censorship cannot easily compensate for information manipulation.

Other important factors for censorship resiliency are the demand for the information being concealed, and the ability to pay the costs to circumvent censorship.

---

[30] Court of Justice of the European Union. Judgment in Case C-131/12 *Google Spain SL, Google Inc. v Agencia Española de Protección de Datos, Mario Costeja González*. PRESS RELEASE No 70/14. Luxembourg, 13 May 2014.

[31] Court of Justice of the European Union. Judgment in Case C-131/12 *Google Spain SL, Google Inc. v Agencia Española de Protección de Datos, Mario Costeja González*. PRESS RELEASE No 70/14. Luxembourg, 13 May 2014.

Entertainment content is more resilient to online censorship than political content, and users with more education, technology access, and wider, more diverse social networks are more resilient to censorship attempts.

4.3.3 *Digital Divide*

Despite its tremendous growth, Internet access is not distributed equally within or between countries. The digital divide refers to "the gap between people with effective access to information and communications technology (ICT), and those with very limited or no access". The gap between people with Internet access and those without is one of many aspects of the digital divide. Whether someone has access to the Internet can depend greatly on their financial status, geographical location as well as government policies. "Low-income, rural, and minority populations have received special scrutiny as the technological 'have-nots'".

Government policies play a tremendous role in bringing Internet access to or limiting access for underserved groups, regions, and countries. Access to computers is a dominant factor in determining the level of Internet access. Internet access has changed the way in which many people think and has become an integral part of people's economic, political, and social lives. The United Nations has recognised that providing Internet access to more people in the world will allow them to take advantage of the "political, social, economic, educational, and career opportunities" available over the Internet. Several of the 67 principles adopted at the World Summit on the Information Society, convened by the United Nations in Geneva in 2003, directly address the digital divide.[32]

To promote economic development and a reduction of the digital divide, national broadband plans have been and are being developed to increase the availability of affordable high-speed Internet access throughout the world.

4.3.3.1 Examples

In Pakistan, the number of Internet users grew from 133,900 (0.1% of the population) in 2000 to 31 million (17.6% of the population) in 2011. Pakistan is now pursuing an aggressive IT policy aimed at boosting its drive for economic modernisation.[33]

---

[32] https://www.un.org/press/en/2003/pi1549.doc.htm

[33] Annan Saeed Khan, Atta ur Rahman, Laila Taskeen Qazi. (2016) "The Relationship Between Internet Usage, Socioeconomic Status, Subjective Health and Social Status". *Business & Economic Review*: Vol. 8, Special Edition: 2016 pp. 67–82.

In North Korea, there is relatively little access to the Internet due to the government's fear of political instability that might accompany the benefits of access to the global Internet.[34]

4.3.4 Digital Rights

Digital rights are those human rights and legal rights that allow individuals to access, use, create, and publish digital media or to access and use computers, other electronic devices, and telecommunications networks. The concept is particularly related to the protection and realisation of existing rights, such as the right to privacy and freedom of expression, in the context of digital technologies, especially the Internet. The laws of several countries recognise a right to Internet access. A number of human rights have been identified as relevant with regard to the Internet. These include freedom of expression, privacy, and freedom of association. Furthermore, the rights to education and multilingualism, consumer rights, and capacity building in the context of the right to development have also been identified.

The APC Internet Rights Charter was established by the Association for Progressive Communications (APC) at the APC Europe Internet Rights Workshop, held in Prague, February 2001.[35] The Charter draws on the People's Communications Charter and develops seven themes: Internet access for all; freedom of expression and association; access to knowledge, shared learning and creation—free and open source software and technology development; privacy, surveillance, and encryption; governance of the Internet; awareness, protection, and realisation of rights. The APC states that "the ability to share information and communicate freely using the Internet is vital to the realisation of human rights as enshrined in the Universal Declaration of Human Rights, the International Covenant on Economic, Social and Cultural Rights, the International Covenant on Civil and Political Rights, and the Convention on the Elimination of All Forms of Discrimination against Women".[36]

In December 2003, the World Summit on the Information Society (WSIS) was convened under the auspices of the United Nations (UN).[37] After lengthy negotiations between governments, businesses, and civil society representatives, the WSIS Declaration of Principles was adopted, reaffirming human rights[38]:

[34] King, Robert R. "North Koreans Want External Information, But Kim Jong-Un Seeks to Limit Access".

Center for Strategic and International Studies (CSIS). May 15, 2019. Available at: https://www.csis.org/analysis/north-koreans-want-external-information-kim-jong-un-seeks-limit-access

[35] https://www.apc.org/en/pubs/about-apc/apc-internet-rights-charter

[36] https://www.apc.org/sites/default/files/APC_charter_EN_0_1_2.pdf

[37] https://www.un.org/press/en/2003/pi1550.doc.htm

[38] https://www.un.org/press/en/2003/pi1549.doc.htm

> We reaffirm the universality, indivisibility, interdependence, and interrelation of all human rights and fundamental freedoms, including the right to development, as enshrined in the Vienna Declaration. We also reaffirm that democracy, sustainable development, and respect for human rights and fundamental freedoms as well as good governance at all levels are interdependent and mutually reinforcing. We further resolve to strengthen the rule of law in international as in national affairs.

The WSIS Declaration also makes specific reference to the importance of the right to freedom of expression in the "Information Society" in stating:

> We reaffirm, as an essential foundation of the Information Society, and as outlined in Article 19 of the Universal Declaration of Human Rights, that everyone has the right to freedom of opinion and expression; that this right includes freedom to hold opinions without interference and to seek, receive and impart information and ideas through any media and regardless of frontiers. Communication is a fundamental social process, a basic human need and the foundation of all social organisation. It is central to the Information Society. Everyone, everywhere should have the opportunity to participate and no one should be excluded from the benefits the Information Society offers.[39]

The 2004 WSIS Declaration of Principles also acknowledged that "it is necessary to prevent the use of information resources and technologies for criminal and terrorist purposes, while respecting human rights".[40]

The Dynamic Coalition for an Internet Bill of Rights held a large preparatory Dialogue Forum on Internet Rights in Rome in September 2007 and presented its ideas at the Internet Governance Forum (IGF) in Rio in November 2007, leading to a joint declaration on Internet rights.[41]

On October 29, 2008, the Global Network Initiative (GNI) was founded upon its "Principles on Freedom of Expression and Privacy".[42] The Initiative was launched in the 60th Anniversary year of the Universal Declaration of Human Rights (UDHR) and is based on internationally recognised laws and standards for human rights on freedom of expression and privacy set out in the UDHR, the *International Covenant on Civil and Political Rights* (ICCPR)[43] and the *International Covenant on Economic, Social and Cultural Rights* (ICESCR). Participants in the Initiative include the Electronic Frontier Foundation, Human Rights Watch, Google, Microsoft, Yahoo, other major companies, human rights NGOs, investors, and academics.

---

[39] Peter Leuprecht, Brave New Digital World? Reflections on the World Summit on the Information Society, 2003 16–1 *Revue québécoise de droit international* 101, 2003 CanLIIDocs 226.

[40] https://www.un.org/press/en/2003/pi1550.doc.htm

[41] https://www.intgovforum.org/multilingual/content/dynamic-coalition-on-internet-rights-and-principles-irpc

[42] Reporters Without Borders, "Why Reporters Without Borders Is Not Endorsing the Global Principles on Freedom of Expression and Privacy for ICT Companies Operating in InternetRestricting Countries," October 28, 2008, http://www.rsf.org/article.php3?id_article=29117

[43] Office of the High Commissioner, United Nations Human Rights. *International Covenant on Civil and Political Rights*, General Assembly resolution 2200A (XXI). December 16, 1966.

https://www.ohchr.org/en/instruments-mechanisms/instruments/international-covenant-civil-and-political-rights

A poll of 27,973 adults in 26 countries, including 14,306 Internet users, was conducted for the BBC World Service by the international polling firm GlobeScan using telephone and in-person interviews between 30 November 2009 and 7 February 2010.[44] GlobeScan Chairman Doug Miller felt, overall, that the poll showed that:

Despite worries about privacy and fraud, people around the world see access to the Internet as their fundamental right. They think the web is a force for good, and most don't want governments to regulate it. Findings from the poll include:

- Nearly four in five (78%) Internet users felt that the Internet had brought them greater freedom.
- Most Internet users (53%) felt that "the Internet should never be regulated by any level of government anywhere".
- Opinion was evenly split between Internet users who felt that "the Internet is a safe place to express my opinions" (48%) and those who disagreed (49%).
- The aspects of the Internet that cause the most concern include fraud (32%), violent and explicit content (27%), threats to privacy (20%), state censorship of content (6%), and the extent of corporate presence (3%).
- Almost four in five Internet users and non-users around the world felt that access to the Internet was a fundamental right (50% strongly agreed, 29% somewhat agreed, 9% somewhat disagreed, 6% strongly disagreed, and 6% had no opinion).

The 88 recommendations made by the Special Rapporteur on the promotion and protection of the right to freedom of opinion and expression in a May 2011 report to the Human Rights Council of the United Nations General Assembly include several that bear on the question of Internet access[45]:

> 67. Unlike any other medium, the Internet enables individuals to seek, receive and impart information and ideas of all kinds instantaneously and inexpensively across national borders. By vastly expanding the capacity of individuals to enjoy their right to freedom of opinion and expression, which is an "enabler" of other human rights, the Internet boosts economic, social and political development, and contributes to the progress of humankind as a whole. In this regard, the Special Rapporteur encourages other Special Procedures mandate holders to engage on the issue of the Internet with respect to their mandates.
>
> 78. While blocking and filtering measures deny users access to specific content on the Internet, States have also taken measures to cut off access to the Internet entirely. The Special Rapporteur considers cutting off users from Internet access, regardless of the justification provided, including on the grounds of violating intellectual property rights law, to be disproportionate and thus a violation of article 19, paragraph 3, of the International Covenant on Civil and Political Rights.
>
> 79. The Special Rapporteur calls upon all States to ensure that Internet access is maintained at all times, including during times of political unrest.

---

[44] https://globescan.com/2021/11/05/new-global-poll-shows-growing-public-support-for-carbon-tax/

[45] https://www.ohchr.org/en/special-procedures/sr-freedom-of-opinion-and-expression

> 85. Given that the Internet has become an indispensable tool for realising a range of human rights, combating inequality, and accelerating development and human progress, ensuring universal access to the Internet should be a priority for all States. Each State should thus develop a concrete and effective policy, in consultation with individuals from all sections of society, including the private sector and relevant Government ministries, to make the Internet widely available, accessible and affordable to all segments of population.

These recommendations have led to the suggestion that Internet access itself is or should become a fundamental human right.[46]

4.3.5 Freedom of Information

Freedom of information is the freedom of a person or people to publish and consume information. Access to information is the ability for an individual to seek, receive, and impart information effectively. This sometimes includes "scientific, indigenous, and traditional knowledge; freedom of information, building of open knowledge resources, including open Internet and open standards, and open access and availability of data; preservation of digital heritage; respect for cultural and linguistic diversity, such as fostering access to local content in accessible languages; quality education for all, including lifelong and e-learning; diffusion of new media and information literacy and skills, and social inclusion online, including addressing inequalities based on skills, education, gender, age, race, ethnicity, and accessibility by those with disabilities; and the development of connectivity and affordable ICTs, including mobile, the Internet, and broadband infrastructures".

Freedom of information (or information freedom) also extends to include the protection of the right to freedom of expression with regard to the Internet and information technology. Freedom of information may also concern censorship in an information technology context, i.e. the ability to access Web content, without censorship or restrictions.

Public access to government information, including through the open publication of information, and formal freedom of information laws, is widely considered an important basic component of democracy and integrity in government.

Michael Buckland—an emeritus professor at the UC Berkeley School of Information and co-director of the Electronic Cultural Atlas Initiative—defines six types of barriers that have to be overcome in order for access to information to be achieved[47]:

---

[46] https://www.ohchr.org/en/HRBodies/HRC/RegularSessions/Session29/Documents/A.HRC.29.32_AEV.doc

[47] Buckland, M. (2011). Data management as bibliography. *Bulletin of the American Society for Information Science and Technology*, 37(6), 34–37. https://doi.org/10.1002/bult.2011.1720370611 Retrieved from https://escholarship.org/uc/item/30w017cq

1. Identification of the source
2. Availability of the source
3. Price of the user
4. Cost to the provider
5. Cognitive access
6. Acceptability

While "access to information", "right to information", "right to know" and "freedom of information" are sometimes used as synonyms, the diverse terminology does highlight particular (albeit related) dimensions of the issue.

Freedom of information is related to freedom of expression, which can apply to any medium, be it oral, written, print, electronic, or through art forms. This means that the protection of freedom of speech as a right includes not only the content but also the means of expression. Freedom of information is a separate concept which sometimes comes into conflict with the right to privacy in the content of the Internet and information technology. As with the right to freedom of expression, the right to privacy is a recognised human right and freedom of information acts as an extension to this right. The government of the United Kingdom has theorised it as being an extension of freedom of speech and a fundamental human right.[48] It is recognised in international law.

Access to information faces great difficulties because of the global digital divide. A digital divide is an economic and social inequality with regard to access to, use of, or impact of information and communication technologies (ICT). The divide within countries (such as the digital divide in the United States) may refer to inequalities between individuals, households, businesses, or geographic areas, usually at different socioeconomic levels or other demographic categories. The divide between differing countries or regions of the world is referred to as the global digital divide, examining this technological gap between developing and developed countries on an international scale.

4.3.5.1 Racial Divide

Although many groups in society are affected by a lack of access to computers or the Internet, communities of colour are specifically observed to be negatively affected by the digital divide. A report published by the DC-based public interest group Fress Press, "Digital Denied: The Impact of Systemic Racial Discrimination on Home-Internet Adoption", concluded that structural barriers and discrimination that perpetuates bias against people of different races and ethnicities contribute to having an impact on the digital divide.[49] The report also concludes that those who do not have

[48] "What is the Freedom of Information Act?". *ico.org.uk*. 4 April 2019. https://ico.org.uk/for-organisations/guide-to-freedom-of-information/what-is-the-foi-act/

[49] https://www.freepress.net/sites/default/files/legacy-policy/digital_denied_free_press_report_december_2016.pdf

Internet access still have a high demand for it, and reduction in the price of home Internet access would allow for an increase in equitable participation and improve Internet adoption by marginalised groups.

Digital censorship and algorithmic bias are observed to be present in the racial divide. Hate-speech rules as well as hate speech algorithms online platforms such as Facebook have favoured white males and those belonging to elite groups in society over marginalised groups in society, such as women and people of colour. In a collection of internal documents that were collected in a project conducted by ProPublica, Facebook's guidelines in regards to distinguishing hate speech and recognising protected groups revealed slides that identified three groups, each one containing either female drivers, black children, or white men. When the question of which subset group is protected was presented, the correct answer was white men. Minority group language is negatively impacted by automated tools of hate detection due to human bias that ultimately decides what is considered hate speech and what is not.

Online platforms have also been observed to tolerate hateful content directed towards people of colour but restrict content from people of colour. Aboriginal memes on a Facebook page were posted with racially abusive content and comments depicting Aboriginal people as inferior. While the contents on the page were removed by the originators after an investigation conducted by the Australian Communications and Media Authority, Facebook did not delete the page and has allowed it to remain under the classification of controversial humour. However, a post by an African American woman addressing her discomfort with being the only person of colour in a small-town restaurant was met with racist and hateful messages. When reporting the online abuse to Facebook, her account was suspended by Facebook for three days for posting the screenshots, while those responsible for the racist comments she received were not suspended. Shared experiences between people of colour can be at risk of being silenced under removal policies for online platforms.

4.3.5.2 Disability Divide

Inequities in access to information technologies are present among individuals living with a disability in comparison to those who are not living with a disability. According to The Pew Internet, 54% of households with a person who has a disability have home Internet access compared to 81% of households that have home Internet access but do not have a person who has a disability.[50] The type of disability an individual has can prevent one from interacting with computer screens and smartphone screens, such as having a quadriplegic disability or having a disability in their hands. However, there is still a lack of access to technology and home Internet access among those who have a cognitive and auditory disability as

[50] https://www.pewresearch.org/internet/2012/04/13/digital-differences/

well. There is a concern about whether or not the increase in the use of information technologies will increase equality through offering opportunities to individuals living with disabilities or whether it will only add to the present inequalities and lead to individuals living with disabilities being left behind in society. Issues such as the perception of disabilities in society, Federal and state government policy, corporate policy, mainstream computing technologies, and real-time online communication have been found to contribute to the impact of the digital divide on individuals with disabilities.

Online abuse towards individuals with disabilities is a factor that can discourage people from engaging online, which could prevent people from learning information that could improve their lives. Many individuals living with disabilities face online abuse in the form of accusations of benefit fraud and "faking" their disability for financial gain, which in some cases leads to unnecessary investigations.

4.3.5.3 Gender Divide

Women's freedom of information and access to information globally is less than that of men.[51] Social barriers such as illiteracy and lack of digital empowerment have created stark inequalities in navigating the tools used for accessing information, often exacerbating a lack of awareness of issues that directly relate to women and gender, such as sexual health. There have also been examples of more extreme measures, such as local community authorities banning or restricting mobile phone use for girls and unmarried women in their communities. According to the Wharton School of Public Policy, the expansion of Information and Communication technology (ICT) has resulted in multiple disparities that have had an impact on women's access to ICT with the gender gap being as high as 31% in some developing countries and 12% globally in 2016.[52] Socioeconomic barriers that result from these disparities are known as what we call the digital divide.

Among low-income countries and low-income regions alike, the high price of Internet access presents a barrier to women since they are generally paid less and face an unequal dividend between paid and unpaid work. Cultural norms in certain countries may prohibit women from having access to the Internet and technology as well by preventing women from attaining a certain level of education or from being the breadwinners in their households, thus resulting in a lack of control over household finances. However, even when women have access to ICT, the digital divide is still prevalent.

---

[51] https://wekh.ca/wp-content/uploads/2021/03/Canadian_Women_Entrepreneurs.pdf

[52] https://documents1.worldbank.org/curated/en/861491551113547855/pdf/134846-WP-PUBLIC-march-2-WB-Women-Study-EN.pdf

4.3.5.4 Sexuality Divide

The Internet provides information that can create a safe space for marginalised groups such as the LGBTQIA+ community to connect with others and engage in honest dialogues and conversations that are affecting their communities. It can also be viewed as an agent of change for the LGBTQIA+ community and provide a means of engaging in social justice. It can allow for LGBTQIA+ individuals who may be living in rural areas or in areas where they are isolated to gain access to information that is not within their rural system as well as gain information from other LGBT+ individuals. This includes information such as health care, partners, and news. GayHealth provides online medical and health information, and the Gay and Lesbian Alliance Against Defamation contains online publications and news that focus on human rights campaigns and issues focused on LGBTQIA issues. The Internet also allows LGBTQIA individuals to maintain anonymity.

Lack of access to the Internet can hinder these things, due to a lack of broadband access in remote rural areas. LGBT Tech has emphasised launching newer technologies with 5G technology in order to help close the digital divide that can cause members of the LGBTQIA community to lose access to reliable and fast technology that can provide information on health care, economic opportunities, and safe communities.

There are other factors that can prevent LGBTQIA+ members from accessing information online or subject them to having their information abused. Internet filters are also used to censor and restrict LGBTQIA+ content that is in relation to the LGBTQIA+ community in public schools and libraries. There is also the presence of online abuse by online predators that target LGBTQIA members by seeking out their personal information and providing them with inaccurate information. The use of the Internet can provide a way for LGBTQIA individuals to gain access to information to deal with societal setbacks through therapeutic advice, social support systems, and an online environment that fosters a collaboration of ideas and concerns and helps LGBTQIA individuals move forward.

4.3.6 Net Neutrality and the Open Internet

Network neutrality, most called net neutrality, is the principle that Internet service providers (ISPs) must treat all Internet communications equally, and not charge users different rates based on content, website, platform, application, type of equipment, source address, destination address, or method of communication. Network neutrality is the principle that all Internet traffic should be treated equally. Internet traffic includes all of the different messages, files, and data sent over the Internet, including emails, digital audio files, digital video files, torrents, etc.

With net neutrality, ISPs may not intentionally block, slow down, or charge money for specific online content. Without net neutrality, ISPs may prioritise certain types of traffic, metre others, or potentially block traffic from specific services, while charging consumers for various tiers of service.

The term was coined by Columbia University media law professor Tim Wu in 2003,[53] as an extension of the longstanding concept of a common carrier, which was used to describe the role of telephone systems. Net neutrality regulations may be referred to as "common carrier" regulations. Net neutrality does not block all abilities that Internet service providers have to impact their customers' services. Opt-in/opt-out services exist on the end user side, and filtering can be done on a local basis, as in the filtering of sensitive material for minors. According to Tim Wu, a public information network will be most useful if all content, websites, and platforms (e.g., mobile devices, video game consoles, etc.) are treated equally, which is the principle of network neutrality.

Net neutrality is the principle that an ISP has to provide access to all sites, content, and applications at the same speed, under the same conditions without blocking or giving preference to any content. Under net neutrality, whether you connect to Netflix, the Internet Archive, or a friend's blog, your ISP has to treat them all the same. Without net neutrality, an ISP can decide what information you are exposed to. Proponents argue that this could cause an increase in monetary charges for companies such as Netflix in order to stream their content.

Research suggests that a combination of policy instruments will help realise the range of valued political and economic objectives central to the network neutrality debate. Combined with public opinion, this has led some governments to regulate broadband Internet services as a public utility, similar to the way electricity, gas, and water supply are regulated, along with limiting providers and regulating the options those providers can offer.

Proponents of net neutrality, which include computer science experts, consumer advocates, human rights organisations, and Internet content providers, assert that net neutrality helps to provide freedom of information exchange, promotes competition and innovation for Internet services, and upholds standardisation of Internet data transmission which was essential for its growth. Opponents of net neutrality, which include ISPs, computer hardware manufacturers, economists, technologists, and telecom equipment manufacturers, argue that net neutrality requirements would reduce their incentive to build out the Internet, would reduce competition in the marketplace, and may raise their operating costs, which they would have to pass along to their users.

Net neutrality is administered on a national or regional basis, though much of the world's focus has been on the conflict over net neutrality in the United States. Under an "open Internet" system, the full resources of the Internet and the means to operate on them should be easily accessible to all individuals, companies, and organisations. Applicable concepts include:

---

[53]Tim Wu, *Network Neutrality, Broadband Discrimination*, Journal of Telecommunications and High Technology Law, Vol. 2, p. 141, 2003 (2003). Available at: https://scholarship.law.columbia.edu/faculty_scholarship/1281

1. Net neutrality
2. Open standards
3. Transparency
4. Lack of Internet censorship
5. Low barriers to entry

The concept of the open Internet is sometimes expressed as an expectation of decentralised technological power and is seen by some observers as closely related to open-source software, a type of software program whose maker allows users access to the code that runs the program, so that users can improve the software or fix bugs.

Proponents of net neutrality see this as an important component of an "open Internet", wherein policies such as equal treatment of data and open web standards allow those using the Internet to easily communicate and conduct business and activities without interference from a third party.

In contrast, a "closed Internet" refers to the opposite situation, wherein established persons, corporations, or governments favour certain uses, restrict access to necessary web standards, artificially degrade some services, or explicitly filter out content. Some countries block certain websites or types of sites, and monitor and/or censor Internet use using Internet police, a specialised type of law enforcement, or secret police.[54] In comparison to the United States or Canada, for example, these countries have far more restrictive Internet service providers. This approach is reminiscent of a closed platform system, as both ideas are highly similar. These systems all serve to hinder access to a wide variety of Internet services, which is a stark contrast to the idea of an open Internet system.

Legal enforcement of net neutrality principles takes a variety of forms, from provisions that outlaw anti-competitive blocking and "throttling" of Internet services, all the way to legal enforcement that prevents companies from subsidising Internet use on particular sites. Contrary to popular rhetoric and statements by various individuals involved in the ongoing academic debate, research suggests that a single policy instrument (such as a no-blocking policy or a quality-of-service tiering policy) cannot achieve the range of valued political and economic objectives central to the debate. As Bauer and Obar suggest, "safeguarding multiple goals requires a combination of instruments that will likely involve government and nongovernment measures. Furthermore, promoting goals such as the freedom of speech, political participation, investment, and innovation calls for complementary policies".[55]

---

[54] Access Now. *The return of digital authoritarianism: internet shutdowns in 2021.* https://www.accessnow.org/keepiton-2021-report

[55] Bauer, Johannes M.; Obar, Jonathan A. *Reconciling Political and Economic Goals in the Net Neutrality Debate.* 2014. The Information Society.

4.3.6.1 United States

Net neutrality in the United States has been a point of conflict between network users and service providers since the 1990s.[56] Much of the conflict over net neutrality arises from how Internet services are classified by the Federal Communications Commission (FCC) under the authority of the *Communications Act of 1934*. The FCC would have significant ability to regulate ISPs should Internet services be treated as a Title II "common carrier service", or otherwise the ISPs would be mostly unrestricted by the FCC if Internet services fell under Title I "information services".[57,58]

In 2009, the United States Congress passed the *American Recovery and Reinvestment Act 2009*,[59] which granted a stimulus of $2.88 billion for extending broadband services into certain areas of the United States.[60] It was intended to make the Internet more accessible for underserved areas, and aspects of net neutrality and open access were written into the grant. However, the bill never set any significant precedents for net neutrality or influenced future legislation relating to net neutrality.[61]

Through 2017, the FCC has generally been favourable towards net neutrality, treating ISPs under Title II common carriers.[62] With the onset of the Presidency of Donald Trump in 2017, and the appointment of Ajit Pai, an opponent of net neutrality, to the chairman of the FCC,[63] the FCC has reversed many previous net neutrality rulings and reclassified Internet services as Title I information services.[64] The FCC's decisions have been the subject of several ongoing legal challenges by both states supporting net neutrality and ISPs challenging it.[65] The United States

---

[56] Gilroy, Angele A. (April 15, 2019). *The Net Neutrality Debate: Access to Broadband Networks* (PDF). Washington, DC: Congressional Research Service. https://fas.org/sgp/crs/misc/R40616.pdf

[57] *Communications Act of 1934*.

[58] https://docs.fcc.gov/public/attachments/DOC-347927A1.pdf

[59] Long title: "An Act making supplemental appropriations for job preservation and creation, infrastructure investment, energy efficiency and science, assistance to the unemployed, State, and local fiscal stabilization, for the fiscal year ending September 30, 2009, and for other purposes."

[60] https://www.govinfo.gov/content/pkg/PLAW-111publ5/html/PLAW-111publ5.htm

[61] "Estimated Impact of the American Recovery and Reinvestment Act on Employment and Economic Output from October 2011 Through December 2011" (PDF). *A CBO Report*. Congressional Budget Office. February 2012. http://www.cbo.gov/sites/default/files/cbofiles/attachments/02-22-ARRA.pdf

[62] "The Net Neutrality Debate All On One Page". *Tech Crunch*. https://techcrunch.com/2008/08/31/the-net-neutrality-debate-all-on-one-page/

[63] https://www.nytimes.com/2017/02/05/technology/trumps-fcc-quickly-targets-net-neutrality-rules.html

[64] Federal Communications Commission. "The Open Internet" (PDF). *FCC Consumer Facts*. http://www.fcc.gov/cgb/consumerfacts/openinternet.pdf

[65] Federal Communications Commission. "Preserving the Open Internet" (PDF). *Policy Statement*. https://apps.fcc.gov/edocs_public/attachmatch/FCC-10-201A1.pdf

Congress has attempted to pass legislation supporting net neutrality but has failed to gain sufficient support.

In 2018, a bill cleared the U.S. Senate, with Republicans Lisa Murkowski, John Kennedy, and Susan Collins joining all 49 Democrats, but the House majority denied the bill a hearing.[66] Individual states have been trying to pass legislation to make net neutrality a requirement within their state, overriding the FCC's decision. California has successfully passed its own net neutrality act, which the United States Department of Justice challenged on a legal basis.

On February 8, 2021, the U.S. Justice Department withdrew its challenge to California's data protection law. Federal Communications Commission Acting Chairwoman Jessica Rosenworcel voiced support for an open Internet and restoring net neutrality. This is an ongoing debate.

In the United States, the federal government and governmental agencies are bound by the *Privacy Act of 1974*[67] and the *Federal Information Security Management Act* (FISMA). [68] Additionally, the United States has sector-specific law which applies to health care as well as banking and finance.

The *Health Insurance Portability and Accountability Act of 1996* (HIPAA) is a US federal law that requires the creation of national standards to protect sensitive patient health information from being disclosed without the patient's consent or knowledge. This was amended to include the Privacy Rule and the Security Rule to enable implementation of the requirements of HIPAA.[69]

Similarly, in finance, the *Gramm–Leach–Bliley Act* (GLBA), also known as the *Financial Services Modernization Act of 1999*, requires a wide range of financial institutions to adequately explain their information-sharing practices to their customers and to safeguard sensitive personal data. The three major components overlaying the collection, disclosure, and protection of consumers' personal information in the *Gramm–Leach–Bliley Act* include: the Financial Privacy Rule, the Safeguards Rule, and Protection from Pretexting.[70]

4.3.6.2 Canada

Net neutrality in Canada is a debated issue in that nation, but not to the degree of partisanship in other nations, such as the United States, in part because of its federal regulatory structure and pre-existing supportive laws that were enacted decades

[66] https://www.politico.com/story/2018/05/16/senate-democrats-net-neutrality-539275

[67] *Privacy Act of 1974*, 5 USC § 552a.

[68] *Federal Information Security Management Act of 2002*, 44 USC § 3541.

[69] *Health Insurance Portability and Accountability Act*. Pub. L. No. 104–191, § 264, 110 Stat.1936.

[70] *Gramm–Leach–Bliley Act* (GLBA), 15 USC. § 6801 et seq.

before the debate arose.[71] In Canada, Internet service providers (ISPs) generally provide Internet service in a neutral manner. Some notable incidents otherwise have included Bell Canada's throttling of certain protocols and Telus's censorship of a specific website supporting striking union members.[72] On October 22, 2009, the CRTC issued a ruling about Internet traffic management, which favoured adopting guidelines that were suggested by interest groups such as OpenMedia.ca and the Open Internet Coalition. However, the guidelines set in place require citizens to file formal complaints proving that their Internet traffic is being throttled, and as a result, some ISPs still continue to throttle the Internet traffic of their users.[73]

Since 2021, the Government of Canada and the CRTC have been working on creating new legislation which intends to address some of the issues of censorship and net neutrality. This legislation has not yet been passed by the House of Commons but is under review.

4.3.6.3 India

In the year 2018, the Indian Government unanimously approved new regulations supporting net neutrality.[74] The regulations are considered to be the "world's strongest" net neutrality rules, guaranteeing free and open Internet for nearly half a billion people, and are expected to help the culture of start-ups and innovation. The only exceptions to the rules are new and emerging services like autonomous driving and telemedicine, which may require prioritised Internet lanes and faster than normal speeds.[75]

4.3.6.4 China

Net Neutrality in China is not enforced, and ISPs in China play important roles in regulating the content that is available domestically on the Internet. There are several ISPs filtering and blocking content at the national level, preventing domestic Internet users from accessing certain sites or services or foreign Internet users from gaining

---

[71] Braga, Matthew (14 December 2017). "Why Canada's net neutrality fight hasn't been as fierce as the one in the U.S." CBC. CBC News.http://www.cbc.ca/news/technology/us-canada-net-neutrality-party-politics-fcc-crtc-fight-1.4447558

[72] "Telus cuts subscriber access to pro-union website". *CBC News*. 24 July 2005. https://www.cbc.ca/news/canada/telus-cuts-subscriber-access-to-pro-union-website-1.531166

[73] Anderson, Steve (2014). *Dynamic Fair Dealing*. University of Toronto Press. pp. 133–143.

[74] Iyengar, Rishi (12 July 2018). "India now has the 'world's strongest' net neutrality rules". *Cnn.com*. https://money.cnn.com/2018/07/12/technology/india-net-neutrality-rules-telecom/index.html

[75] Doval, Pankaj (12 July 2018). "Web stays equal for all as govt clears net neutrality". *The Times of India*. https://timesofindia.indiatimes.com/business/india-business/internet-to-remain-free-and-fair-in-india-govt-approves-net-neutrality/articleshow/64948838.cms

access to domestic web content. This filtering technology is referred to as the Great Firewall, or GFW.[76]

In an article published by the Cambridge University Press, they observed the political environment of net neutrality in China.[77] Chinese ISPs have become a way for the country to control and restrict information rather than provide neutral Internet content for those who use the Internet. Unlike countries such as the United States of America, values such as free speech and equality are not even in question when it comes to China's standards of net neutrality.

4.3.7 Privacy

Internet privacy involves the right or mandate of personal privacy concerning the storing, repurposing, provision to third parties, and displaying of information pertaining to oneself via the Internet. Internet privacy is a subset of data privacy. Privacy concerns have been articulated from the beginnings of large-scale computer sharing.

Privacy can entail either personally identifiable information (PII) or non-PII information such as a site visitor's behaviour on a website. PII refers to any information that can be used to identify an individual. For example, age and physical address alone could identify who an individual is without explicitly disclosing their name, as these two factors are unique enough to identify a specific person typically. Other forms of PII may soon include GPS tracking data used by apps, as the daily commute and routine information can be enough to identify an individual.

It has been suggested that the "appeal of online services is to broadcast personal information on purpose".[78] On the other hand, in his essay "The Value of Privacy", security expert Bruce Schneier says, "Privacy protects us from abuses by those in power, even if we're doing nothing wrong at the time of surveillance".[79,80]

---

[76] https://www.websiterating.com/vpn/glossary/what-is-great-firewall-china/

[77] Cambridge University Press (September 2011). "The Political Economy of Governing ISPs in China". *The China Quarterly*. 207: 523–540. https://www.cambridge.org/core/journals/china-quarterly/article/abs/the-political-economy-of-governing-isps-in-china-perspectives-of-net-neutrality-and-vertical-integration/678FA3FEDDF28AE8B096CE9FC743B6F5

[78] Pogue, David (January 2011). "Don't Worry about Who's watching". *Scientific American*. **304** (1): 32.

[79] "The Value of Privacy by Bruce Schneier". Schneier.com. https://www.schneier.com/blog/archives/2006/05/the_value_of_pr.html

[80] Bruce Schneier (May 18, 2006). "The Eternal Value of Privacy by Bruce Schneier". Wired.com. http://archive.wired.com/politics/security/commentary/securitymatters/2006/05/70886

4.3.7.1 Privacy Levels

Internet and digital privacy are viewed differently from traditional expectations of privacy. Internet privacy is primarily concerned with protecting user information. Law Professor Jerry Kang explains that the term privacy expresses space, decision, and information.[81] In terms of space, individuals have an expectation that their physical spaces (e.g., homes, cars) not be intruded. Lastly, information privacy is in regard to the collection of user information from a variety of sources, which produces great discussion.

At the end of the 1990s, with the rise of the Internet, it became clear that governments, companies, and other organisations would need to abide by new rules to protect individuals' privacy. With the rise of the Internet and mobile networks, Internet privacy is a daily concern for users.

People with only a casual concern for Internet privacy need not achieve total anonymity. Internet users may protect their privacy through the controlled disclosure of personal information. The revelation of IP addresses, non-personally identifiable profiling, and similar information might become acceptable trade-offs for the convenience that users could otherwise lose using the workarounds needed to suppress such details rigorously. On the other hand, some people desire much stronger privacy. In that case, they may try to achieve Internet anonymity to ensure privacy—use of the Internet without giving any third parties the ability to link the Internet activities to personally-identifiable information of the Internet user. In order to keep their information private, people need to be careful with what they submit to and look at online. When filling out forms and buying merchandise, information is tracked, and because it was not private, some companies send Internet users spam and advertising for similar products.

Posting things on the Internet can be harmful or expose people to malicious attacks. Some information posted on the Internet persists for decades, depending on the terms of service and privacy policies of particular services offered online. This can include comments written on blogs, pictures, and websites, such as Facebook and Twitter. It is absorbed into cyberspace, and once it is posted, anyone can potentially find it and access it. Some employers may research a potential employee by searching online for the details of their online behaviours, possibly affecting the outcome of the success of the candidate.

4.3.7.2 Privacy Risks

Companies are hired to track which websites people visit and then use the information, for instance, by sending advertising based on one's web browsing history. There are many ways in which people can divulge their personal information, for

[81] Kang, Jerry. (2004). Information Privacy in Cyberspace Transactions. Stanford Law Review. 50(4). 10.2307/1229286.

instance, through the use of "social media" and by sending bank and credit card information to various websites. Moreover, directly observed behaviour, such as browsing logs, search queries, or contents of the Facebook profile, can be automatically processed to infer potentially more intrusive details about an individual, such as sexual orientation, political and religious views, race, substance use, intelligence, and personality.

Those concerned about Internet privacy often cite a number of privacy risks—events that can compromise privacy—which may be encountered through online activities. These range from the gathering of statistics on users to more malicious acts such as the spreading of spyware and the exploitation of various forms of bugs—or software faults.

Several social networking websites try to protect the personal information of their subscribers, as well as provide a warning through a privacy and terms agreement. On Facebook, for example, privacy settings are available to all registered users: they can block certain individuals from seeing their profile, they can choose their "friends", and they can limit who has access to their pictures and videos. Privacy settings are also available on other social networking websites such as Google Plus and Twitter. The user can apply such settings when providing personal information on the Internet. The Electronic Frontier Foundation has created a set of guides so that users may more easily use these privacy settings, and Zebra Crossing, an easy-to-use digital safety checklist, is a volunteer-maintained online resource.[82]

In late 2007, Facebook launched the Beacon program in which user rental records were released to the public for friends to see.[83] Many people were enraged by this breach of privacy, and the Lane v. Facebook, Inc. case ensued.[84]

4.3.7.3 Risks for Young People

Children and adolescents often use the Internet (including social media) in ways that risk their privacy: a cause for growing concern among parents. Young people also may not realise that all their information and browsing can and may be tracked while visiting a particular site and that it is up to them to protect their own privacy. They must be informed about all these risks. For example, on Twitter, threats may include shortened links that may lead to potentially harmful websites or content. Email threats include email scams and attachments that persuade users to install malware and disclose personal information. On torrent sites, threats include malware hidden in video, music, and software downloads. When using a smartphone, threats include geolocation, meaning that one's phone can detect one's location and post it online for all to see. Users can protect themselves by updating virus protection, using security settings, downloading patches, installing a firewall, screening email,

[82] "Protecting Yourself on Social Networks". *Surveillance Self-Defense*. 2014-09-29.

[83] https://circleid.com/posts/3973/6034/

[84] https://circleid.com/posts/3973/6034/

shutting down spyware, controlling cookies, using encryption, fending off browser hijackers, and blocking pop-ups.

However, many people have little idea how to go about doing these things. Many businesses hire professionals to take care of these issues, but most individuals can only do their best to educate themselves.

4.3.7.4 Big Data

Big data is generally defined as the rapid accumulation and compiling of massive amounts of information that is being exchanged over digital communication systems. The data is large (often exceeding exabytes) and cannot be handled by conventional computer processors and is instead stored on large server-system databases. This information is assessed by analytic scientists using software programs, which paraphrase this information into multi-layered user trends and demographics. This information is collected from all around the Internet, such as by popular services like Facebook, Google, Apple, Spotify, or GPS systems. Big data provides companies with the ability to[85]:

- Infer detailed psycho-demographic profiles of Internet users, even if they were not directly expressed or indicated by users.
- Inspect product availability and optimise prices for maximum profit while clearing inventory.
- Swiftly reconfigure risk portfolios in minutes and understand future opportunities to mitigate risk.
- Mine customer data for insight, and create advertising strategies for customer acquisition and retention.
- Identify customers who matter the most.
- Create retail coupons based on a proportional scale to how much the customer has spent, to ensure a higher redemption rate.
- Send tailored recommendations to mobile devices at just the right time, while customers are in the right location to take advantage of offers.
- Analyse data from social media to detect new market trends and changes in demand.
- Use clickstream analysis and data mining to detect fraudulent behaviour.
- Determine root causes of failures, issues, and defects by investigating user sessions, network logs, and machine sensors.

[85] Meru University College of Science and Technology. "Privacy issues of medical applications."

4.3.7.5 Other Privacy Risks

Internet users obtain Internet access through an Internet service provider (ISP). All data transmitted to and from users must pass through the ISP. Thus, an ISP has the potential to observe users' activities on the Internet. ISPs can breach personal information such as transaction history, search history, and social media profiles of users. Hackers could use this opportunity to hack ISPs and obtain sensitive information from victims.

However, ISPs are usually prohibited from participating in such activities for legal, ethical, business, or technical reasons.

Normally ISPs do collect at least some information about the consumers using their services. From a privacy standpoint, ISPs would ideally collect only as much information as they require in order to provide Internet connectivity (IP address, billing information if applicable, etc.).

Which information an ISP collects, what it does with that information, and whether it informs its consumers pose significant privacy issues. Beyond the usage of collected information typical of third parties, ISPs sometimes state that they will make their information available to government authorities upon request. In the United States and other countries, such a request does not necessarily require a warrant.

An ISP cannot know the contents of properly encrypted data passing between its consumers and the Internet. For encrypting web traffic, https has become the most popular and best-supported standard. Even if users encrypt the data, the ISP still knows the IP addresses of the sender and the recipient. (However, see the IP addresses section for workarounds.)

An anonymizer such as I2P—The Anonymous Network or Tor [86]—can be used for accessing web services without them knowing one's IP address and without one's ISP knowing what the services are that one accesses. Additional software has been developed that may provide more secure and anonymous alternatives to other applications. For example, Bitmessage can be used as an alternative to email and Cryptocat as an alternative to online chat. On the other hand, in addition to End-to-End encryption software, there are web services such as Qlink which provide privacy through a novel security protocol which does not require installing any software.

While signing up for Internet services, each computer contains a unique IP, Internet Protocol address. This particular address will not give away private or personal information, however, a weak link could potentially reveal information from one's ISP. The following list presents some of the methods by which information from one's ISP can be gathered and used:

[86] "Characterization of Tor Traffic Using Time Based Features", Arash Habibi Lashkari, Gerard Draper-Gil, Mohammad Saiful Islam Mamun and Ali A. Ghorbani, In the proceeding of the third International Conference on Information System Security and Privacy, SCITEPRESS, Porto, Portugal.

- Cross-device tracking identifies users' activity across multiple devices.
- Massive personal data extraction through mobile device apps that receive carte-blanche permissions for data access upon installation.
- Malware is a term short for "malicious software" and is used to describe software to cause damage to a single computer, server, or computer network, whether that is through the use of a virus, Trojan Horse, spyware, etc.
- Spyware is a piece of software that obtains information from a user's computer without that user's consent.
- A web bug is an object embedded into a web page or email and is usually invisible to the user of the website or reader of the email. It allows checking to see if a person has looked at a particular website or read a specific email message.
- Phishing is a criminally fraudulent process of trying to obtain sensitive information such as usernames, passwords, and credit card or bank information. Phishing is an Internet crime in which someone masquerades as a trustworthy entity in some form of electronic communication.
- Pharming is a hacker's attempt to redirect traffic from a legitimate website to a completely different Internet address. Pharming can be conducted by changing the hosts file on a victim's computer or by exploiting a vulnerability on the DNS server.
- Social engineering where people are manipulated or tricked into performing actions or divulging confidential information.
- Malicious proxy server or other "anonymity" services.
- Use of weak passwords that are short, consist of all numbers, all lowercase or all uppercase letters, or that can be easily guessed such as single words, common phrases, a person's name, a pet's name, the name of a place, an address, a phone number, a social security number, or a birth date.
- Use of recycled passwords or the same password across multiple platforms which have become exposed due to a data breach.
- Using the same login name and/or password for multiple accounts where one compromised account leads to other accounts being compromised.
- Allowing unused or little used accounts, where unauthorised use is likely to go unnoticed, to remain active.
- Using out-of-date software that may contain vulnerabilities that have been fixed in newer, more up-to-date versions.
- WebRTC is a protocol which suffers from a serious security flaw that compromises the privacy of VPN tunnels, by allowing the true IP address of the user to be read. It is enabled by default in major browsers such as Firefox and Google Chrome.

4.4 Resurgence of Sovereignty

As the underlying political concepts—and the frameworks, beliefs, and concepts which guide political decision-making—pertaining to the Internet and online activities move to the forefront of digital awareness, the architecture of the Internet has also shifted, in the direction of great sovereign control of networks and network activities. One of the major risks of this expansion of sovereignty to the online world is the concern around the fragmentation of governance, in which the underlying protocols support global connectivity on paper but, in practice, present with many fractured, uncoordinated, and highly dissonant rules for data, privacy, and security. With concerns about privacy and security growing among the public and private sectors, governments have been compelled to play a greater role in the protection of citizens.

We can consider the main ideological frameworks to be based around three practical concerns: protection, preservation, and prevention. These concerns align with the items discussed in the following section, in which we examine three key issues in data sovereignty:

1. Encryption and the hybrid cloud;
2. Data localisation; and,
3. Data breach prevention.

4.4.1 Encryption and the Hybrid Cloud

Digital collaboration and its inherent data-sharing are a fact of life within the modern, cloud-connected enterprise. The traditional assumption of trust given to appropriately credentialed employees and contractors operating within the enterprise's hardened perimeter has been extended to modern cloud-based architectures. Teams working together to advance business objectives can freely share data across private hybrid cloud, multi-cloud, and on-premises-based shared data storage.

All of these activities are founded upon data-sharing across a tenuous ecosystem of trust that could be strengthened through encryption. Unfortunately, encryption is often underutilised—if utilised at all—due to the compartmentalisation of technologies that exist to protect data in its various states and the increased user friction at both ends of the transaction. To make matters worse, if a data recipient breaks confidence and shares the data with others—whether on purpose or by accident—the data owner/custodian has no knowledge of the breach of trust unless the offending party informs them. Enterprise data must be better, and more broadly, protected via a holistic, highly integrated set of technologies. This will aid the data owners/custodians in maintaining control and tracking their data throughout its lifecycle.

4.4.1.1 Hybrid Cloud Security

Hybrid Cloud Security can include encrypting data in transmission, at rest/in storage, as well as in processing and use.

Encrypting Data in Transmission

The proper application of encryption can significantly reduce the losses from data theft. There are pros and cons to using transport-based encryption throughout the internal environment. The main benefit to those who own the data is the higher level of confidentiality it brings when they are moving data for legitimate business purposes. The main problem for those who own the data is the higher level of confidentiality the encryption brings to black hats when they are moving stolen data. Security monitoring is often blind to what is in the traffic, or they must invest in gateway proxy tools to intercept traffic with an authorised man-in-the-middle attack to see what it is. This type of inspection can also add latency to applications, creating other problems and making the choice to use transport-based encryption difficult and often expensive. It is also interesting to note that even before the transport encryption boom, the number of successful attacks on transport encryption was low, and the number of records stolen from data at rest was high. Stealing data on the fly is a more complex task that requires more technical setup and timeliness, especially to steal targeted data, than the technical setup for stealing data at rest.

Encrypting Data at Rest/in Storage

Data at rest is by far the largest theft target in hybrid cloud. Historically, encryption systems have proved to be costly and difficult to install, configure, operate, and maintain. Business users who interact with the data criticise the friction it brings to their work and the negative performance impacts it has on their work, even to the point of causing failed/lost requests. Thus, in the battle of usability versus security, usability is still winning. For storage, conquering these issues requires addressing the underlying usability of the crypto system by focusing on crypto-key management. The usability of the data encryption tools to remove friction from the data customers, whether internal employees or external clients, is also important.

Encrypting Data in Processing and Use

Protecting data as it flows through an application (data in process) is probably the most difficult aspect of data risk control. The first control gate is normal access provisioning. If someone does not have access to the front door of the application, then accessing the data is far more difficult. Beyond that, the focus of control is more often on physically protecting servers or hardening the electronic components that comprise the system. Attacks on data in processing require direct access to the system for probing, direct access to the data being operated on, or malware inserted into the application or system drivers to funnel data out while the application is operating. Depending on how processing is structured, controls may need to be added to defend against people syphoning data as they are handling/processing it (data in use). Caution should be taken to ensure that the right data goes to the right person. Traditionally, this is where encryption has been deployed the most. So long as only the proper people were included in the circle of trust, the data owner or custodian could be confident that the data was safe. However, until recently, the crucial limitation was that once the data left the owner or custodian's possession, it was still possible for someone to decrypt the information and forward it to another party unbeknownst to the data owner/custodian.

4.4.1.2 Transparency

Transparent means that end-users are unaware of the encryption/decryption processes, and end-to-end means that data is encrypted at-rest and in-transit. Transparent encryption provides protection for data at rest. When transparent encryption is applied, the protection is removed before data is accessed. For example, when an authorised user copies a file from a file server, this not only makes the encryption process "transparent" to end users but also means data exists in the clear any time it is moved or copied from the protected location. In contrast, persistent encryption is encryption that travels with data as it is shared, copied, and moved from one system or user to another.

4.4.1.3 Hybridisation

According to Webster's, a hybrid is something "having or produced by a combination of two or more distinct elements" that is "marked by heterogeneity in origin". In other words, it's a blended version of different pieces that is often designed to perform the same tasks as the originals but does so in a different or more flexible way. Inherent in the concept of hybrid, from a technology perspective, is that existing solutions or technologies can only solve or provide the capabilities for a portion of a given problem or task. It takes the combination of multiple pieces to create a more complete solution. **Hybridisation** in cloud computing refers to a division of processing power, data storage, mission-critical applications, and customer interaction between internal, private, and externally managed hybrid clouds. Most of the cloud service providers are enjoying the benefits of migrating or hybridisation in the infrastructure for cost reduction, disaster recovery compliance, or expanded service offering even though this process is both timely and essential.

4.4.2 Data Localisation

In addition to having an impact on governing bodies, data sovereignty laws also have serious implications for corporations. With the rise of social media, and the integration of cloud-services into our daily lives, there have been a number of scandals and controversies regarding non-consensual data sharing and/or breaches. We have also seen some of the significant issues which arise from conflicting data sovereignty laws and have resulted in some of the ongoing national—and regional—law reviews with respect to cross-border data sharing.

On March 17, 2018, just weeks prior to the implementation of the General Data Protection Regulation by the European Union, the New York Times[87] and the Guardian[88] broke the news that 50 million Facebook user profiles had been harvested for "psychographic profiles" by Cambridge Analytica. Facebook later reported that the number of Facebook user profiles, from which information was mined for targeted political campaigns and other projects, was over 87 million.[89] On March 26, 2018, just more than a week after the initial story was published, Facebook stock fell by about 24%, equivalent to $134 billion.[90]

Just less than two months later, in May of 2018, Wall Street reported that Facebook had been able to recover its losses.[91] While the impact on Facebook was mostly overcome, this scandal was able to shine a stark light of reality on the fragility of our assumptions of personal data protection in an online world. Unfortunately, while the Facebook-Cambridge Analytica scandal brought data and its importance to the forefront of consumer awareness, this level and scope of public awareness was only able to be reached after the personal data of millions of users was harvested through a massive breach of trust.

In another controversy, the Court of Justice of the European Union (CJEU) has recently determined that cloud-based services hosted in the United States are incapable of complying with the European Union's *General Data Protection Regulation* and other EU privacy laws.[92] In their decision, the Court of Justice of the European Union established that data exports are not merely financial decisions, as the fundamental rights of the users must also be considered as a matter of priority. This decision specifically impacts large-scale global data operators, such as Google Analytics and the Meta corporation.

[87] https://www.nytimes.com/2018/03/17/us/politics/cambridge-analytica-trump-campaign.html

[88] https://www.theguardian.com/news/2018/mar/17/cambridge-analytica-facebook-influence-us-election

[89] https://www.nytimes.com/2018/04/04/technology/mark-zuckerberg-testify-congress.html

[90] https://www.cbsnews.com/news/facebook-stock-price-recovers-all-134-billion-lost-in-after-cambridge-analytica-datascandal/

[91] https://www.bloomberg.com/news/articles/2021-10-06/facebook-stock-s-familiar-crisis-cycle-decline-rebound-repeat

[92] https://curia.europa.eu/juris/document/document.jsf?text=&docid=228728&pageIndex=0&doclang=EN&mode=req&dir=&occ=first&part=1&cid=12078024

In response, Meta issued a stark warning to Europe in their annual report filed with the Securities and Exchange Commission (SEC) which suggests that Facebook and Instagram will no longer be available within the European Union unless data is allowed to flow to its servers in the United States.[93] The company argues that the lack of an acceptable data transfer framework "would materially and adversely affect [the] business, financial condition, and results of operations" as "Meta, and many other businesses, organisations, and services, rely on data transfers between the EU and the US in order to operate global services".[94]

> If a new transatlantic data transfer framework is not adopted and we are unable to continue to rely on SCCs (standard contractual clauses) or rely upon other alternative means of data transfers from Europe to the United States, we will likely be unable to offer a number of our most significant products and services, including Facebook and Instagram, in Europe.—Meta Annual Report.[95]

The outcomes of the ruling made by the Court of Justice of the European Union could impact many corporations, both in the United States and in the European Union. On March 25, 2022, following extensive media coverage, the European Commission and the United States announced that they have agreed in principle on a new Trans-Atlantic Data Privacy Framework.[96] The joint statement describes the framework as providing "a durable basis for trans-Atlantic data flows, which are critical to protecting citizens' rights and enabling trans-Atlantic commerce in all sectors of the economy, including for small and medium enterprises" and promoting "an inclusive digital economy in which all people can participate and in which companies of all sizes from all of our countries can thrive".

4.4.3 Breach Prevention, Remediation, and Planning

In the field of information technology, there are two main facets which deal with data breaches: data breach prevention and data breach remediation. When these facets are combined and working in tandem, this is known as data breach planning.

---

[93] https://d18rn0p25nwr6d.cloudfront.net/CIK-0001326801/14039b47-2e2f-4054-9dc5-71bcc7cf01ce.pdf

[94] https://d18rn0p25nwr6d.cloudfront.net/CIK-0001326801/14039b47-2e2f-4054-9dc5-71bcc7cf01ce.pdf

[95] https://d18rn0p25nwr6d.cloudfront.net/CIK-0001326801/14039b47-2e2f-4054-9dc5-71bcc7cf01ce.pdf

[96] https://www.whitehouse.gov/briefing-room/statements-releases/2022/03/25/fact-sheet-united-states-and-european-commission-announce-trans-atlantic-data-privacy-framework/

4.4.3.1 Data Breach Prevention

Companies who experienced breaches in the past skimped on encrypted storage and multifactor authentication security measures to save time and increase productivity. Round-the-clock, real-time monitoring is essential to preventing data breaches. Companies can only prevent data breaches if they have visibility into their environment. Outsourcing to a security operations centre to monitor activity 24/7 can be a good option, along with automated intrusion detection and prevention solutions. Tools like a security information and event management platform (SIEM) provide actionable insights from advanced analytics and data forensics to mitigate risks and speed up your incident response. SIEM systems consolidate logs from devices throughout your network to detect patterns in data over time; this intelligence allows you to categorise issues according to their severity, so you can prioritise actions and make informed security decisions. Employee expertise is vital, and security personnel should receive regular training so that they are aware of policies, expectations, and data security standards. As IT technology advances, so too should employee training so that staff are equipped with the latest tools and know how to apply them. Training is an investment that will pay dividends in the end.

4.4.3.2 Data Breach Remediation

If a breach happens, there are certain steps that can mitigate and contain an incident. Companies must do everything in their power to protect customers and shareholders and be transparent about their efforts to build trust. After a breach, a company should stem the flow and stop additional data loss by identifying vulnerabilities and fixing them using several strategies including:

Organise an Expert Response Team

A response team should include forensic, legal, IT, HR, operations, communications, investor relations staff, and management experts. This team will be in charge of dealing with the aftermath of a breach in all aspects of your business.

Secure the Physical Area

Stemming the loss requires securing the physical area as well as systems. Equipment should be taken offline or replaced (depending on the situation), access to your logical and physical environment should be monitored and frozen as necessary, and passwords and access codes for authorised users should be updated.

Check Websites and Online Sources

Any misplaced information should be removed from public sources. Internet search engines can cache information, but you can directly contact a search engine to make sure they do not archive personal information. Other websites should be contacted and asked to remove any exposed data that they may have obtained from a compromised site.

Remove Vulnerabilities

Check the network segmentation with forensic experts and see who gained access to what using the data from your SIEM agents and logs, and you may uncover the source of the issue. Investigate whether or not the proper procedures and technology were used and review your access logs for clues. When identifying vulnerabilities, consider that they may originate from your third-party providers. Verify which data, applications, and systems were compromised prior to taking action, and determine the number of individuals affected. Next, you'll want to contact stakeholders to let them know a security incident has occurred and what steps you're taking to recover.

Create a Communication Plan

Now is the time for transparency and communications both internally and externally. Companies that try to bury the extent of their breach, often experience much worse consequences in terms of customer backlash. Consider all stakeholders in the plan, including business partners, investors, employees, and customers. Deliver a clear statement that is open and thorough without divulging information that could add risk. Organisations should report all data breaches to law enforcement and follow other state and federal legal requirements—including compliance mandates.

4.5 Global Complexity

While the creation and maintenance of domestic laws can be challenging, when those needs are expanded to an international scale, the global implications make the creation and maintenance of such laws significantly more complex.

In this section, we will explore the global issues pertaining to Internet standards, network latency, cyberbalkanization, Internet shutdowns, media freedom, and privacy and encryption.

4.5.1 Global Internet Standards

The Internet, a loosely organised international collaboration of autonomous, interconnected networks, supports host-to-host communication through voluntary adherence to open protocols and procedures defined by Internet Standards. There are also many isolated interconnected networks, which are not connected to the global Internet but use the Internet standards. An Internet standard (STD) is a specification that has been approved by the Internet Engineering Task Force (IETF). In general, an Internet Standard is a specification that is stable and well-understood, is technically competent, has multiple, independent, and interoperable implementations with substantial operational experience, enjoys significant public support, and is recognizably useful in some or all parts of the Internet. Such a standard helps to promote a consistent and universal use of the Internet worldwide. Prior to approval, the proposed Internet standard goes through a series of stages called a "standards track". The standard first starts as a draft, which may evolve into an RFC (Request for Comments). If the IETF approves the RFC, then it becomes a standard. RFCs and proposed Internet standards cover a lot of ground. From standards for handling technical TCP/IP processes to standards for the display of various types of media, the IETF weighs in on a large number of these standards as they make their way toward adoption. We have two common types of Internet standards:

4.5.1.1 Web Standards

TCP/IP model and associated Internet standards Web standards are a type of Internet standard which defines aspects of the World Wide Web. They allow for the building and rendering of websites. The three key standards used by the World Wide Web are Hypertext Transfer Protocol (HTTP), and Uniform Resource Locator (URL). Respectively, they specify the content and layout of a web page, what web page identifiers mean, and the transfer of data between a browser and a web server.

4.5.1.2 Network Standards

Network standards are a type of Internet standard which defines rules for data communication in networking technologies and processes. Internet standards allow for the communication procedure of a device to or from other devices. In reference to the TCP/IP Model, common standards and protocols in each layer are as follows: transport layer (TCP[97] and SPX[98]), network layer (IP[99] and IPX[100]), data link layer

[97] https://en.wikipedia.org/wiki/Internet_protocol_suite
[98] https://en.wikipedia.org/wiki/IPX/SPX
[99] https://en.wikipedia.org/wiki/Internet_protocol_suite
[100] https://en.wikipedia.org/wiki/Internetwork_Packet_Exchange

(IEEE 802.3[101] for LAN and Frame Relay[102] for WAN), and physical layer (8P8C[103] and V.92[104]).

4.5.1.3 Compliance Standards

There are a wide variety of different standards compliance reports which can be made available to customers. By addressing their data residency requirements with local regulations and securely and by privately localising regulated data, such as Payment Card Information (PCI), Protected Health Information (PHI), and Personally Identifiable Information (PII), companies can expand their business into new countries. There are several standards for compliance. Some of these include SOC 1 Type II, SOC 2 Type II, SOC 3, GxP, PCI DSS, HIPAA, 152-FZ, CSA, ISO 27001, ISO27017, ISO27018, ISO 27701, and the EU's *General Data Protection Regulation* (GDPR).[105] We can describe those in greater detail.

- ***SOC 1 Type II***
  Internal controls report is specifically intended to meet the needs of the customers, management, and their auditors, as they evaluate the effect of security controls on their own internal controls for financial reporting. The SOC 1 report examination is performed in accordance with the Statement on Standards for Attestation Engagements (SSAE) No. 16 and the International Standard on Assurance Engagements (ISAE) No. 3402; therefore it can be used by customers and their auditors in both the United States and abroad. These reports are issued by independent third-party auditors periodically.
- ***SOC 2 Type II***
  A SOC 2 Type II (Service Organisation Controls) report is based on the AICPA Trust Service Principles (TSP) and Criteria and covers the security, availability, and confidentiality of TSPs as they relate to a Cloud Service Provider (CSP).
- ***SOC 3***
  A SOC 3 (Service Organization Controls) report is an abbreviated version of a SOC 2 report and is appropriate for users who want assurance about the Cloud Service Provider's (CSP) controls but do not require a full SOC 2 report. A SOC 3 report may only be issued if the CSP has an unqualified audit.
- ***GxP***

---

[101] https://en.wikipedia.org/wiki/IEEE_802.3

[102] https://en.wikipedia.org/wiki/Frame_Relay

[103] https://en.wikipedia.org/wiki/Modular_connector

[104] https://en.wikipedia.org/wiki/V.92

[105] *EU General Data Protection Regulation (GDPR):* Regulation (EU) 2016/679 of the European Parliament and of the Council of 27 April 2016 on the protection of natural persons with regard to the processing of personal data and on the free movement of such data, and repealing Directive 95/46/EC (General Data Protection Regulation), OJ 2016 L 119/1.

Good Clinical Practices, Good Laboratory Practices, and Good Manufacturing Practices ("GxP") Compliance on platforms enables a secure and highly available data distribution and localisation, aligned to the requirements of life science organisations for validated and controlled workloads. Customers will benefit from an improved user experience, reduced costs, and improved security.

- ***PCI DSS***

  The Payment Card Industry-Data Security Standard (PCI-DSS) is an industry-level information security standard regarding the secure handling of credit card information.
- ***HIPAA***

  The Health Insurance Portability and Accountability Act (HIPAA) provides U.S. federal protections for patient health information held by covered entities (CEs) and business associates (BAs).
- ***152-FZ***

  Russia's Federal Law No. 152 and meets industry standards as well as satisfying the requirements under the law "On Personal Data".
- ***CSA***

  The Security Trust Assurance and Risk (STAR) Level 2 Certification is a rigorous third-party independent assessment of the security of a cloud service provider. The certification leverages the requirements of the ISO/IEC 27001:2013 management system standard together with the CSA Cloud Controls Matrix criteria.
- ***ISO 27001***

  ISO/IEC 27001:2013 is a risk-based set of information security requirements that require an organisation to have a well-structured information security management system (ISMS). Maintenance of the system requires annual audits by external auditors, ongoing risk assessments, and continuous improvement of the system.
- ***ISO27017***

  ISO/IEC 27017:2015 provides guidance on the information security aspects of cloud computing, recommending the implementation of cloud-specific information security controls that supplement the guidance of the ISO/IEC 27002 and ISO/IEC 27001 standards. This code of practice provides additional information security control implementation guidance specific to cloud service providers.
- ***ISO27018***

  ISO/IEC 27018:2019 is a code of practice that focuses on the protection of personal data in the cloud. It is based on ISO/IEC information security standard 27,002 and provides implementation guidance on ISO/IEC 27002 controls applicable to public cloud Personally Identifiable Information (PII).
- ***ISO 27701***

  The design goal of ISO 27701 is to enhance the existing information security management system (ISMS) with additional requirements in order to establish, implement, maintain, and continually improve a Privacy Information Management System (PIMS). The standard outlines a framework for Personally

Identifiable Information (PII) Controllers and PII Processors to manage privacy controls to reduce the risk to the privacy rights of individuals.

- ***EU General Data Protection Regulation***

  The European Union (EU) passed the General Data Protection Regulation (GDPR) to give individuals more control over their personal data. GDPR imposes more rules about processing PII and comes with powerful enforcement. We invest significant strategic resources in maintaining compliance with the GDPR and we also aim to help our customers comply with the processes and policies outlined.[106]

4.5.2 Network Latency for Data Transmission

Network latency, sometimes called **lag**, is the term used to describe delays in communication over a network. Latency in networking is best thought of as the amount of time it takes for a packet of data to be captured, transmitted, processed through multiple devices, then received at its destination and decoded. When delays in transmission are small, it's referred to as a low-latency network (desirable), and longer delays are called a high-latency network (not so desirable). Long delays that occur in high-latency networks create bottlenecks in communication. In the worst cases, it's like traffic on a four-lane highway trying to merge into a single lane. High latency decreases communication bandwidth and can be temporary or permanent, depending on the source of the delays. There are several reasons for network latency, including[107]:

- **Distance**

  One of the main causes of network latency is distance, or how far away the device making requests is located from the servers responding to those requests.
- **Website Construction**

  The way web pages are constructed makes a difference in latency. Web pages that carry heavy content, large images, or load content from several third-party websites may perform more slowly, as browsers need to download larger files to display them.
- **End-user Issues**

  Network problems might appear to be responsible for latency, but sometimes RTT latency is the result of the end-user device being low on memory or CPU cycles to respond in a reasonable timeframe.
- **Physical Issues**

[106] *EU General Data Protection Regulation (GDPR):* Regulation (EU) 2016/679 of the European Parliament and of the Council of 27 April 2016 on the protection of natural persons with regard to the processing of personal data and on the free movement of such data, and repealing Directive 95/46/EC (General Data Protection Regulation), OJ 2016 L 119/1.

[107] https://www.ir.com/guides/what-is-network-latency

In a physical context, common network latency causes are the components that move data from one point to the next—physical cabling such as routers, switches, and Wi-Fi access points. In addition, latency can be influenced by other network devices like application load balancers, security devices, firewalls, and Intrusion Prevention Systems (IPS).[108]

4.5.3 Cyberbalkanization

Cyberbalkanization is the segregation of the Internet into smaller groups with similar interests, to the extent that they show a narrow-minded approach to outsiders or those with contradictory views. While the Internet has largely been credited for broadening discussion, it also can serve as a means of bringing together fringe groups with intolerant viewpoints. So, while the Internet has contributed to globalisation and information exchange, it also may be used to foster discrimination. This phenomenon, also more casually known as the "splinternet," whereby governments seek to fence off the World Wide Web into a series of national Internets, is not new. The term, also known as cyberbalkanization, has been around since the 1990s.

The term "cyberbalkanization" combines the term "cyber" with the Balkans, a political region in southeastern Europe with a history of partitioned cultures, languages, and religions. One of the key advantages of the Internet is that it has allowed all kinds of people to publicise their views and reach out to others with similar interests. As a result, however, it also has the same power to bring together those with outlying—and often discriminatory—viewpoints. The ability to communicate allows such communities to flourish more than they would if these people were unable to communicate because of geographical separation. As a result, the Internet also has served to promote and incite hatred against minority groups, as well as promote political or social agendas that do not find much footing in popular media.

Splintering of the Internet community can occur when people engage in confirmation bias and create echo chambers, using the Internet to exclude or avoid views that contradict their beliefs and attitudes. Called cyberbalkanization (or sometimes cyberbalkanization), it refers to the division of the Internet into sub-groups with specific interests (digital tribes), where the sub-group's members almost always use the Internet or the web to communicate or read material that is only of interest to the rest of the sub-group. The term may have first been used in an MIT paper by Marshall Van Alstyne and Erik Brynjolfsson that was published in late 1996. The concept was also discussed in a related article in the journal Science that same year. The term is a hybrid of cyber, relating to the Internet, and Balkanization, a phenomenon that takes its name from the Balkans, a part of Europe that was historically subdivided by languages, religions, and cultures.

[108] https://www.ir.com/guides/what-is-network-latency

In his 2001 book *Republic.com*, Cass Sunstein argued that cyberbalkanization could damage democracy, because it allows different groups to avoid exposure to one another as they gather in increasingly segregated communities, making recognition of other points of view or common ground decreasingly likely. The commentator Aleks Krotoski feels that Jihadist groups often use the Internet in this way.[109]

Despite the concerns of cyberbalkanization, there is mixed evidence that it is growing. One Wharton study found that Internet filters can create commonality, not fragmentation. However, this study primarily focused on music recommendation algorithms and openly states that more research is required surrounding other domains (e.g. news, books, fashion). Another study found that the ideological segregation of online news consumption is low in absolute terms, higher than the segregation of most offline news consumption, and significantly lower than the segregation of face-to-face interactions with neighbours, co-workers, or family members. The study notes that an important caveat, however, is that none of their evidence speaks to the way people translate the content they encounter into beliefs, which may be a larger factor in the problem these types of studies seek to address.

4.5.4 Internet Shutdowns

Internet shutdowns refer to when state authorities deliberately shut down the Internet. In other cases, Internet shutdown could describe intentional acts by state authorities to slow down Internet connections. Other terms used to describe Internet shutdowns include "blanket shutdowns", "kill switches", "blackouts", and "digital curfews". Shutdowns could be for only a few hours or days, up to weeks or sometimes months.

Governments often justify Internet shutdowns on grounds of public safety, prevention of mass hysteria, hate speech, fake news, national security, and sometimes for transparency of an ongoing electioneering process. However, reports indicate that shutdowns are a deliberate attempt at Internet censorship by the governments.[110] Apart from posing great harm to Internet freedom, the shutdown of the Internet harms public health, economies, educational systems, Internet advancements, vulnerable groups, and democratic societies. This is because they impede public communication through the Internet for a while, thereby putting many activities at a standstill.

In the past years, no fewer than 35 countries have experienced Internet shutdowns. According to reports by Access Now, a non-profit digital rights group, 25 countries across the globe experienced government-induced Internet shutdowns 196 times in 2018. In 2019, Access Now reports indicated that 33 countries

---

[109] Sunstein, Cass R. *Republic.com*. 2001. Princeton University Press.

[110] Access Now. *The return of digital authoritarianism: internet shutdowns in 2021*. https://www.accessnow.org/keepiton-2021-report

experienced a government-induced Internet shutdowns 213 times. The 2020 report from the digital rights group implied that 29 countries deliberately shut down their Internet 155 times.[111]

With the growing trend of Internet shutdowns, digital rights groups, including Internet Society, Access Now, #KeepItOn Coalition, and others, have condemned it, noting it is an "infringement on digital rights" of netizens. These groups have also been at the forefront of tracking and reporting shutdowns in real-time as well as analysing its impact on Internet advancement, Internet freedom, and societies.

4.5.5 Media Freedom

Media, freedom of expression, and freedom of information have been long recognised as principles of Internet governance, included in the 2003 Geneva Declaration[112] and 2005 Tunis Commitment of the World Summit on the Information Society (WSIS).[113] Given the cross-border, decentralised nature of the Internet, an enabling environment for media freedom in the digital age requires global multi-stakeholder cooperation and shared respect for human rights. In broad terms, two different visions have been seen to shape global Internet governance debates in recent years: fragmentation versus common principles.

On the one hand, some national governments, particularly in the Central and Eastern European and Asia-Pacific regions, have emphasised state sovereignty as an organising premise of national and global Internet governance. In some regions, data localisation laws—requiring that data be stored, processed and circulated within a given jurisdiction—have been introduced to keep citizens' personal data in the country, both to retain regulatory authority over such data and to strengthen the case for greater jurisdiction. Countries in the Central and Eastern European, Asia-Pacific, and African regions all have legislation requiring some localisation.[114] Data localisation requirements increase the likelihood of multiple standards and the fragmentation of the Internet, limiting the free flow of information and, in some

[111] Access Now. *The return of digital authoritarianism: internet shutdowns in 2021.* https://www.accessnow.org/keepiton-2021-report

[112] United Nations Digital Library. Declaration of Principles "Building the Information Society: A Global Challenge to the New Millennium" (2003). World Summit on the Information Society. Phase 1 (2003: Geneva) (DHLAUTH)609,142. Available at: https://digitallibrary.un.org/search?f1=author&as=1&sf=title&so=a&rm=&m1=p&p1=World%20Summit%20on%20the%20Information%20Society.%20Phase%201%20%282003%20%3A%20Geneva%29%20%28DHLAUTH%29609142&ln=en

[113] Second Phase of the WSIS (16–18 November 2005, Tunis) Tunis Commitment WSIS-05/TUNIS/DOC/7. Available at: https://www.itu.int/net/wsis/tunis/index.html

[114] World trends in freedom of expression and media development: global report 2017/2018. UNESCO Publishing, Oxford University. Feb 5, 2018.

cases increasing the potential for surveillance, which in turn impacts on freedom of expression.[115]

On the other hand, the dominant practice has been towards a unified, universal Internet with broadly shared norms and principles. The NETmundial meeting, held in Brazil in 2014, produced a multistakeholder statement that the "Internet should continue to be a globally coherent, interconnected, stable, unfragmented, scalable, and accessible network-of-networks".[116]

In 2015, UNESCO's General Conference endorsed the concept of Internet Universality and the "ROAM Principles", which state that the Internet should be "(1) Human Rights-based (2) Open, (3) Accessible to all, and (4) Nurtured by Multistakeholder participation".[117] The ROAM Principles combine standards for process (multi-stakeholderism to avoid potential capture of the Internet by a single power centre with corresponding risks), with recommendations about substance (what those principles should be).[118] The fundamental position is for a global Internet where ROAM principles frame regional, national and local diversities. In this context, significant objectives are media freedom, network interoperability, net neutrality, and the free flow of information (minimal barriers to the rights to receive and impart information across borders and any limitations to accord with international standards).

In a study of 30 key initiatives aimed at establishing a bill of rights online during the period between 1999 and 2015, researchers at Harvard's Berkman Klein Centre found that the right to freedom of expression online was protected in more documents (26) than any other right.[119] The UN General Assembly committed itself to multistakeholderism in December 2015 through a resolution extending the WSIS process and IGF mandate for an additional decade. It further underlined the importance of human rights and media-related issues such as the safety of journalists.

Growing support for the multistakeholder model was also observed in the Internet Assigned Numbers Authority (IANA) stewardship transition, in which oversight of the Internet's addressing system shifted from a contract with the United States Department of Commerce to a new private sector entity with new multi-stakeholder accountability mechanisms.[120] Another support of the multistakeholder approach has been the Tallinn Manual 2.0 on the *International Law Applicable to Cyber Operations*, the updated and considerably expanded second edition of the 2013 Tallinn Manual on the International Law Applicable to Cyber Warfare. The annual

---

[115] https://unctad.org/system/files/official-document/der2021_en.pdf

[116] https://document.netmundial.br/1-internet-governance-principles/

[117] https://unesdoc.unesco.org/ark:/48223/pf0000243325

[118] Unesco Internet Study. Internet Universality: A Means Towards Building Knowledge Societies and the Post-2015 Sustainable Development Agenda. Available at: http://wayback.archive-it.org/10611/20170505142323/http://www.unesco.org/new/en/communication-and-information/crosscutting-priorities/unesco-internet-study/internet-universality/

[119] World trends in freedom of expression and media development: global report 2017/2018. UNESCO Publishing, Oxford University. Feb 5, 2018.

[120] https://www.internetsociety.org/iana-transition/

conferences linked to the Budapest Convention on Cybercrime and meetings of the United Nations Group of Governmental Experts on Developments in the Field of Information and Telecommunications in the Context of International Security, mandated by the United Nations General Assembly, have deliberated on norms such as protection of critical infrastructure and the application of international law to cyberspace.[121]

4.5.5.1 Examples

In the period 2012–2016, the African Union passed the Convention on Cyber Security and Personal Data Protection, and the Commonwealth Secretariat adopted the Report of the Working Group of Experts on Cybercrime.

The Economic Community of West African States (ECOWAS) compelled all 15 member states to implement data protection laws and authorities through the adoption of the *Supplementary Act on Personal Data Protection* in 2010.[122] Again in 2011, the ECOWAS adopted a Directive on Fighting Cybercrime to combat growing cybercrime activities in the West African region.[123,124] In response to the growing need for ICT infrastructures, cybersecurity, and increasing cybercrime, the ECOWAS, on 18 January 2021, adopted the regional strategy for cybersecurity and the fight against cybercrime.

In a bid to unify data protection across Europe and give data subjects autonomy over their data, the European Union implemented the General Data Protection Regulation on 25 May 2018. It replaced the insufficient Data Protection Directive of 1995. The EU describes it as the "toughest privacy and security law" globally. Under the GDPR, data subjects have the right to access, rectification, erasure, restriction of processing, profiling, object to automated processing, and data portability.[125]

4.5.6 Privacy and Encryption

Privacy and security online have been of paramount concern to Internet users with growing cybercrime and cyberattacks worldwide. A 2019 poll by Safety Monitor

[121] https://www.un.org/disarmament/wp-content/uploads/2018/04/Civil-Society-2017.pdf

[122] https://www.statewatch.org/media/documents/news/2013/mar/ecowas-dp-act.pdf

[123] https://ccdcoe.org/organisations/ecowas/

[124] http://www.tit.comm.ecowas.int/wp-content/uploads/2015/11/SIGNED_Cybercrime_En.pdf

[125] EU General Data Protection Regulation (GDPR): Regulation (EU) 2016/679 of the European Parliament and of the Council of 27 April 2016 on the protection of natural persons with regard to the processing of personal data and on the free movement of such data, and repealing Directive 95/46/EC.

shows that 13 percent of people aged 15 and above have been victims of cybercrimes such as identity fraud, hacking, and cyberbullying in the Netherlands.[126] INTERPOL recommends using encrypted Internet to stay safe online.[127] Encryption technology serves as a channel to ensuring privacy and security online. It is one of the strongest tools to help Internet users globally stay secure on the Internet, especially in the aspect of data protection. However, criminals leverage the privacy, security, and confidentiality of online encryption technology to perpetrate cybercrimes and sometimes absolved themselves of their legal criminal consequences. It has sparked debates between Internet governors and governments of various countries on whether encryption technology should stay or its use should be stopped.

The UK Government, in May 2021, proposed the *Online Safety Bill*,[128] a new regulatory framework to address cyberattacks and cybercrimes in the United Kingdom, but without a strong encryption technology. This is in a bid to make the United Kingdom the safest place to use the Internet in the world and curb the damaging effect of harmful content shared online, including child pornography. However, the Internet Society argues that a lack of strong encryption exposes Internet users to even greater risks of cyberattacks and cybercrimes, adding that it overrides data protection laws.

4.6 Summary

In this chapter, we discussed some of the major technical challenges in data sovereignty and digital governance. We explored the current challenges in domestic law, as well as the challenges encountered in international law and global Internet standards. Beyond merely legal perspectives, we also considered the ethical, moral, and rights-based arguments relating to access to information, freedom of expression, and digital censorship within our international legal landscape.

In the next chapter, we will explore the concept and realisation of digital governance and its intersection with data sovereignty on both a national and an international level.

The answers to the following questions are provided within this chapter:

1. What are some of the methods of technical and non-technical censorship?
2. Why are larger data vaults a technical challenge in data sovereignty and digital governance?
3. What is metadata and why does it matter?

---

[126] https://longreads.cbs.nl/the-netherlands-in-numbers-2020/what-about-cyber-crime/

[127] https://www.interpol.int/en/News-and-Events/News/2022/New-campaign-highlights-digital-extortion-threats-and-how-to-keep-safe

[128] UK Parliamentary House of Commons, *Online Safety Bill*. Up-to-date information is made available at: https://bills.parliament.uk/bills/3137

4. How does the hybrid cloud differ from other cloud models?
5. What does "cyberbalkanization" refer to in the context of international digital law?
6. Why is it important to consider the freedom of the media in relation to global Internet standards and digital censorship?

Chapter 5
Comparative Legal Strategies

As the international community has evolved to accommodate for the digital landscape, we have seen a variety of different strategies emerge between nations, which aim to address the dual-headed issue of data sovereignty and digital governance. Many jurisdictions are increasingly enacting laws and regulations governing how and where data must be stored either within their respective borders or abroad. What has resulted is a constantly evolving network of rules and restrictions for the location of data. As a result of a lack of definitions in legislation within the Americas region, as well as many authorities using the terms interchangeably there is no definitive definition. However, conceptually we can define data residency as rules governing where a jurisdiction specifies their data must be stored geographically. On the other hand, data localisation can be defined generally as the rules which mandate that data created within a certain jurisdiction must be kept within the jurisdiction.

In order to glean a better understanding of the various strategies used to accommodate political and governance frameworks onto our online realm, we can look to a handful of nations, as a sample of frameworks used to date to address these issues.

5.1 Topic Overview

In this chapter we will examine the different legislative solutions of a sample of countries to compare the outcomes and strategies related to data sovereignty, consent and governance around the world. These will be detailed alphabetically.

M. Lukings, A. Habibi Lashkari, *Understanding Cybersecurity Law in Data Sovereignty and Digital Governance*, Progress in IS,
https://doi.org/10.1007/978-3-031-14264-2_5

5.2 National Data Strategies

Comparatively, we can look at the governance provisions given which aim to protect the various rights of individuals and organisations in a governing state. On a large-scale national level, we can look to specific countries as examples of some of the strategies undertaken by governments around the world in an effort to create a balance between profit and protection within the digital era.

There are common restrictions on certain types of data. Government-related data generally attracts residency requirements across all regions in the Americas. In addition, there are similar trends for accounting and tax data whereby either original copies must be kept within the region, or the transfer of such data must be authorised by relevant authorities, e.g. Argentina, Colombia and Canada.

Most jurisdictions in the Americas—specifically Canada and the United States—also have implicit data residency requirements because of the common restrictions for cross-border transfers of personal data. A number of jurisdictions have adopted a similar approach to that of the European Union's *General Data Protection Regulation* and set out restrictions, including authorisation from authorities, security measures and adequacy requirements for countries where data will be transferred to. From here, we will describe the related legislation for each country, with our sample countries listed alphabetically.

5.2.1 Canada

There is no explicit definition for either data residency or data localisation under Canadian law. Although the *Personal Information Protection and Electronic Documents Act,* SC 2000 c 5 ("PIPEDA"), provides a comprehensive data protection legislation, its data residency obligations focus on how data can be transferred abroad instead of where data should be stored. Notably, there are quite strict localisation rules in Canada both at the federal level and provincial level associated with certain categories of data, e.g. government-related data. This is a notable trend across many jurisdictions as governments look to protect data that could have an impact on national security or public safety.

In Canada, there are different statutory privacy protections which separately apply to the government, to corporations and to individuals. As well, regardless of the category into which an individual, corporation, organisation or other entity falls, all parties to whom Canadian law is applicable must abide by, and operate within, the provisions of the *Canadian Charter of Rights and Freedoms.*[1] The Government of Canada, in their White Paper on *Data Sovereignty and Public Cloud,* outlines the governmental strategy whereby cloud services are identified and evaluated as the

[1] *Part I of the Constitution Act, 1982. Canadian Charter of Rights and Freedoms*, s 7, Part 1 of the *Constitution Act*, 1982, being Schedule B to the Canada Act 1982 (UK), 1982, c 11.

principal delivery option when initiating information technology (IT) investments, initiatives, strategies and projects.[2]

Appendix J: Standard on Security Categorisation of the Directive on Security Management requires that data which has been categorised as either "classified" or "protected" must be kept within Canada or stored in a Government of Canada facility abroad. There are also provincial data residency requirements which provide that data held by certain public sector institutions, such as hospitals, schools and government agencies, must be stored and accessed only in Canada, unless certain conditions are met. We can see specific examples of this in Nova Scotia's *Personal Information International Disclosure Protection Act* and British Columbia's *Freedom of Information and Protection of Privacy Act*.

With respect to the ability of the government to access, collect, use, retain, disclose and destroy data within the federal public sector, the relevant statutory provisions are found in the *Privacy Act*[3] and the *Access to Information Act*.[4] The *Privacy Act*[5] and the *Access to Information Act*[6] were both implemented by the Canadian federal government in 1985 and have acted as a starting point for more recent legislation and privacy laws, including those pertaining to the cyber sector. These Acts work together to provide a legislative framework for personal data collection, use, retention, disclosure and individual access within the federal public sector. For corporations and commercial marketplaces, including private businesses, commercial corporations and similar organisations, they must abide by the provisions established in the *Personal Information Protection and Electronic Documents Act*[7]—commonly, and highly understandably, shortened to "PIPEDA"—and by the rules regulating the use of commercial electronic messages in *Canada's Anti-Spam Legislation*.[8]

Over the past couple of years, there has been some seemingly significant governmental interest shown in creating newer, more broadly applicable, digital online content regulations and mandatory content exposure standards that would specifically apply to Canadian broadcasters operating online.[9] Existing Canadian laws apply both to Canadian companies and to international companies which conduct their business endeavours within Canada. These types of platforms—those which conduct commercial endeavours in Canada—are already subject to well-established

[2] https://www.canada.ca/en/government/system/digital-government/digital-government-innovations/cloud-services/gc-white-paper-data-sovereignty-public-cloud.html

[3] *Privacy Act* (RSC 1985, c P-21).

[4] *Access to Information Act* (RSC 1985, c A-1).

[5] *Privacy Act* (RSC 1985, c P-21).

[6] *Access to Information Act* (RSC 1985, c A-1).

[7] *Personal Information Protection and Electronic Documents Act* (SC 2000, c 5).

[8] *Canada's Anti-Spam Legislation* (SC 2010, c 23).

[9] This would not necessarily be limited to digital content streaming services, but could potentially be extend to include some social networking platforms under the "broadcaster" umbrella.

Canadian law, including the organisational provisions established under the PIPEDA and *Canada's Anti-Spam Legislation,*[10] among other examples.

There is no rule stipulating the federal government must keep its sensitive data in Canada, but the Directive on Digital Service updated in 2020 says keeping computing facilities within borders should be considered as the first choice.[11] There has been some recent political support in Canada for forcing platforms like Facebook, Google, Amazon and Netflix to be subject to Canadian laws and pay Canadian taxes.[12] Support has been shown for a variety of provisions, including those which would require companies which operate in Canada, including via the Internet, to obey the (currently unwritten) laws of Canada's new—but not yet existent—*Digital Charter.* Canada introduced legislation in 2021 that would update its data privacy rules to look more like the GDPR, but the bill never came to pass.[13] Politicians are expected to take another crack at it in 2022 and we have already seen movement on this effort in the first half of the year.[14]

5.2.2 China

Until 2021, China had several laws intersecting and a draft of a prospective data protection law regulating how companies were to handle data. Now that the country's overarching cross-sector law, the *Personal Information Protection Law* (PIPL), has recently passed, the data situation will likely shift to take the law's restrictions into account. From an objective reading of the new law, China will have some of the strictest data protection enforcement in the world.

Data protection in China has advanced more in 2021 than anyone could have expected, with the government choosing data as the next battlefield for control over an increasingly independent-minded tech sphere. The government even issued warnings to some of the largest domestic Chinese companies, issuing clearly that data compliance will be mandatory for all from here on out.

Cross-border data flows are largely not permitted, meaning companies will need to localise data within China. For rare cases where data might go abroad, PIPL outlines a few necessities: First, the Cyberspace Administration of China, the appropriate authority, needs to explicitly approve any data transfer abroad; second, large data handlers and critical information infrastructure providers (CII), which

[10] *Canada's Anti-Spam Legislation* (SC 2010, c 23).

[11] https://www.tbs-sct.gc.ca/pol/doc-eng.aspx?id=32601

[12] See the attempt by the federal liberal government to create new Internet censorship (and more!) regulations in their proposed Bill C-10, which was stalled in the senate prior to the 2021 federal election and which is now undergoing a highly secretive consultation and review process about which the public is (surprisingly?) being given extremely little information.

[13] https://www.csoonline.com/article/3202771/general-data-protection-regulation-gdpr-requirements-deadlines-and-facts.html

[14] https://priv.gc.ca/en/opc-news/news-and-announcements/2020/qa_20200924/

generally constitute any company running a large operation, must localise all data within the country.

In addition to the localisation mandates PIPL contains, other laws also intersect with data protection, most notably the cross-sector Cybersecurity Law 2017 and the payment-specific Announcement of the People's Bank of China No.7 in 2018. Due to the complexity of the layering that governs the issue, we recommend seeking legal consultation from a Chinese law firm.

Companies must get client consent to store and process data, even if the data will remain within China's borders. Under PIPL, fines for noncompliance are massive: up to $7.7 million or 5% of a company's previous year revenues. With China already demanding that domestic companies revise their data handling strategy, often giving them just days to do so, the government will not be hesitant to enforce the new law. Foreign companies should ensure constant compliance to avoid both the significant fines and any corresponding logistical problems.

5.2.3 Egypt

While modelled to a large extent on the EU *General Data Protection Regulation* (GDPR),[15] the law diverges in several key areas, including exempting the Central Bank of Egypt (CBE) and all entities (including banks) subject to its supervision from the scope of the law. Both areas feature prominently in the GDPR. Finally, the GDPR places the principle of transparency (i.e. requiring that any information addressed to individuals be concise, easily accessible, easy to understand and presented in clear and plain language) at its very core. The law does not contain such emphasis.

Article 14 of the law stipulates that in order to transfer personal data out of Egypt, a company must obtain a licence from the regulator and should only transfer personal data to a country which affords the same level of protection to personal data as Egypt under the law. Article 15 sets out a number of exceptions to the Article 14 restriction, which are all contingent on first obtaining the express consent of the data subject. To export personal information from Egypt, a company must first apply for a permit from the yet to be established data protection authority.

The collection and processing of personal data is governed by the newly enacted Egyptian Data Protection Act of 2020. Data protection laws apply to the processing and organising of the data and those who determine the purposes of that processing, the content of the data and any related operations.

[15] *EU General Data Protection Regulation (GDPR):* Regulation (EU) 2016/679 of the European Parliament and of the Council of 27 April 2016 on the protection of natural persons with regard to the processing of personal data and on the free movement of such data and repealing Directive 95/46/EC (General Data Protection Regulation), OJ 2016 L 119/1.

5.2.4 European Union

The European Union has a unified data protection law called the *General Data Protection Regulation* (or GDPR).[16] This law regulates processing of personal data within the European Union and is an important component of the European Union's privacy and human rights law. While the European Union does not currently have specific data localisation requirements, the recent invalidation of Privacy Shield could mean the requirement of such. Many companies have already taken steps to ensure their data strategies ensure data localisation of regulated data before it leaves their countries of origin.

Organizations that receive and hold any regulated data types follow the *General Data Protection Regulation* requirements. According to GDPR, companies have to keep the data secure inside the EU, and if the data is to be transferred outside of the UE, then it can only be transferred to countries or organisations that have signed up to equivalent privacy protection.

How does this work in technical terms? A transfer means that the source data was moved to a machine outside the European Union. But, it can also happen when an employee outside the EU accesses the data—for example, a developer in India is checking logs or a support engineer in Singapore is helping a customer and views his data.

These manipulations are considered as data transfer as well (since data moves to another country), so in an ideal world you'll have to make sure that only EU citizens and UE machines interact with the data. Both storage in, processing in, and access from outside the EEA will count as transfer. This has major implications for your processing architecture. For example, if you have a US and international customer base, you will need to store and process data separately and in multiple countries.

Yet, such transfer is still ok if you agree to apply *General Data Protection Regulation* data protection principles or use a special data residency-as-a-service provider that helps protect the data during transfers.

5.2.5 Germany

Germany's data protections have largely been governed by the European Union's *General Data Protection Regulation* (GDPR) since it was passed in 2016 and enacted in 2018. Germany's data protection landscape is somewhat decentralised however, as each German state has its own data protection authority, with the federal

[16] *EU General Data Protection Regulation (GDPR):* Regulation (EU) 2016/679 of the European Parliament and of the Council of 27 April 2016 on the protection of natural persons with regard to the processing of personal data and on the free movement of such data and repealing Directive 95/46/EC (General Data Protection Regulation), OJ 2016 L 119/1.

authority, the Bundesbeauftragte für den Datenschutz und die Informationsfreiheit (BfDI), only coordinating things on the European Union level.[17]

German state-level law rarely deviates from the *General Data Protection Regulation*, with principles like transparency, data minimisation, client consent, storage limitations and other key components being exactly the same. Also Germany has passed the *Gesetz zur Verbesserung der Rechtsdurchsetzung in sozialen Netzwerken (Netzwerkdurchsetzungsgesetz*—NetzDG law), focused on limiting hate speech on social media.

EU countries—Germany included—have not been shy about fining companies found to have violated the GDPR, particularly regarding violations around client consent.

Many fines are vigorously litigated however, with some companies escaping without pay or seeing the penalty lowered. For this reason, some within the EU have called for the *General Data Protection Regulation* to be tightened to make it easier to enforce on a wide scale.

Cross-border data transfers are permitted under the *General Data Protection Regulation*, as long as the data is going to a country with adequate data protection. For this reason, since the 2020 Schrems II ruling, there have been questions surrounding data flows between EU countries and the United States. As one of the main focus points of the *General Data Protection Regulation*, client consent itself and how it is received are keys to German data protection.

Consent must be freely given, informed and unambiguous and state the specific reasons for data processing. Likewise, individuals can revoke consent at any time. Any violation of any of these factors can open a company up to fines and penalties.

While Germany is and has been one of the leading forces behind European data protections, the country has yet to institute and enact severe penalties like other EU members such as France and Ireland. While *General Data Protection Regulation*[18] fines can be hundreds of millions of dollars, numerous fines handed out by German data protection authorities have been overruled or lowered by German courts. Despite this precedent, there is always the risk of seven-figure fines for noncompliance.

5.2.6 India

India's parliament has been working on the draft of the country's data protection bill for several years. While the draft has been opened up for further debates, the

---

[17] https://www.bfdi.bund.de/EN/Home/home_node.html

[18] *EU General Data Protection Regulation (GDPR):* Regulation (EU) 2016/679 of the European Parliament and of the Council of 27 April 2016 on the protection of natural persons with regard to the processing of personal data and on the free movement of such data and repealing Directive 95/46/EC (General Data Protection Regulation), OJ 2016 L 119/1.

expectation is that it will soon pass and become law. Getting comprehensive data regulations will tremendously benefit the country, as data centre growth within India should skyrocket to match rising demands in a market seen as a sleeping giant.

As India tries to work out a *Data Protection Bill* to finally get through its parliament, it seems like social media companies will be liable to be penalised severely for any potential violations. India has already taken to strictly enforcing existing industry-related regulations, as it has penalised both Mastercard and PayPal for noncompliance in 2021.

Current India data regulations are a patchwork of legislation, with several sectors such as payment and government primarily affected. This is particularly true for payment systems, which the Reserve Bank of India has advised should be stored locally, even if the info can go abroad for short periods of time. This regulation is what has caused Mastercard and PayPal trouble recently.

While there are currently no restrictions banning cross-border data transfers, payment card data and sensitive personal data need to be stored domestically. The comprehensive data protection draft notes that "critical data" cannot leave the country, but does not define the term. Transferring data abroad for processing can occur, but storage of most data should be in India.

5.2.7 Indonesia

Indonesia has managed its data protection through a patchwork of regulations thus far, but its main law, the *Personal Data Protection* bill (PDP), will hopefully conclude and officially be passed in November. This would conclude a 7-year journey for the draft, which started out in 2014 and has undergone notable changes as the government tried to add GDPR-style clauses.

As of now, there are three regulations that have governed data protection throughout the country: GR 71/2019 on government-related information (which in practice covers multiple sectors), Bank Regulation No.18/40/2016, which covers payment and financial services and GR 46/2014 for healthcare.

Without a unifying data protection law, Indonesia has not had much breaking news in the sphere, but the data leak from a now-defunct government healthcare system and the rapid acceleration of data regulations in surrounding Asian countries gives Indonesia a strong impetus to pass a bill in the immediate future. In certain cases, cross-border data transfers are allowed. While the severity of Indonesia's data protection is not extraordinarily high, some information does need to be stored locally, including government and electronic payment information.

Healthcare and other financial data has more flexibility, as banks need to have local data centres for information but can transfer data abroad with a regulator's permission. Healthcare data processing can also be done abroad, if it is accessible to the ministry's domestic database, although it must only be stored domestically, and e-records need to be maintained for at least 10 years. Some of these allowances may change when PDP passes in November, and much of the material that currently exists has not been translated, leaving some uncertainty about the current patchwork approach.

The sector-specific regulations currently in place do not handle the matter of client consent, but the *Personal Data Protection* bill set to debut in November has a heavy focus on offering citizens GDPR-style rights, including client consent, data breach notifications and the right for citizens to request their data be deleted.

Without an overarching data protection bill, penalties for noncompliance don't have a numerical figure on them yet. Instead, violations of sector-specific laws will result in warnings and, eventually, sanctions against the offending organisations.

5.2.8 Japan

Japan's reformed privacy regulation took effect in 2017 reforming the *Act of Personal Information*. The amendment has enabled Japan to be on the European Union's "allowed list" of countries with adequate data protection legislation. Japanese authorities are concerned about where personal information is being transferred overseas. They have issued guidance that companies need to be more careful where data is processed overseas.

There are no registration requirements for the overseas transfer of data and no general data localisation requirements in Japanese laws. Consent must be freely given, informed and state the specific reason, especially for cross-border transfers. Penalties can include criminal and civil actions. Japanese authorities are concerned about where personal information is being transferred overseas. They have issued guidance that companies need to be more careful where data is processed overseas.

5.2.9 Kazakhstan

Data regulations in Kazakhstan are primarily under the Republic of Kazakhstan No. 94-V dated 21 May 2013 *On Personal Data and Its Protection* and the Ministry of Digital Development, Innovation and Aerospace Industry. There are several additional legislative amendments that further the power of the government to enforce data regulations, including *Resolution No. 12* dated 18 January 2021, *On Introducing Amendments to Certain Resolutions of the Government of the Republic of Kazakhstan*.

Similar to its fellow ex-Soviet and CIS representative Russia, data must be stored within the country, although cross-border data transfers are allowed, as processing can occur abroad. This does mean, however, that data processed abroad has to be transferred back to Kazakhstan and exist as a mirrored copy. The level of responsibility put on database owners and operators has increased due to two amendments passed in 2021, and with several typical data protection terms still left undefined—

such as "sensitive data" and "data processor"— there are some grey areas as to exactly how and when the country will enforce the regulations.

Like most countries, Kazakhstan generally requires companies to acquire client consent before collecting personal data. Unlike in stricter countries that demand written consent, electronic consent also suffices in Kazakhstan. While neither the interests of the individual nor public interest carry legal basis, the country has a considerable list of government-related situations when data processing and collection can occur without customer consent.

2020's "Amendments to Certain Legislative Acts of the Republic of Kazakhstan on Informatization Issues" sought to make penalties more severe. While monetary fines for illegally processing or collecting personal data are still quite low, at just $5800 per violation, exposing personal data carries up to 5 years in prison and 7 years if disseminated publically (such as through media).

Since the summer of 2020, Kazakhstan has been passing more and more detailed data protection regulations, bringing the originally sparse law more in line with *General Data Protection Regulation* conventions and Russian data protectionism.

5.2.10 Kuwait

In late 2021, Kuwait passed a comprehensive data protection law, the culmination of years of ramping up its data protections as the entire region aims to transform itself economically using data and tech. With Kuwait and the entire Arabian region trying to shift their economies away from oil and towards FinTech and other modern enterprises, data is now the most valuable resource in Kuwait. Accordingly, the country finally passed comprehensive data protection regulations in 2021.

The governing body, the Communication and Information Technology Regulatory Authority (CITRA), has been more active recently, overseeing *Law No. 20 of 2014* (on electronic transmissions) and two laws passed in 2021: *Cloud Computing Regulatory Framework and Data Privacy Protection Regulation.*

The *Cloud Computing Regulatory Framework* was passed after the primary data regulations to patch some holes and better establish the information and communications technology industry as part of the New Kuwait 2035 project. Companies must receive written client consent before transferring data abroad, in addition to informing individuals why their data is being transferred and who is receiving it. Additionally, rules to protect subscribers have been put in place on cloud service providers against unfair contract terms, including provisions against liability clauses.

There have yet to be published occurrences of any violation of the new laws, but penalties are a maximum of 3 years in prison and a fine of no less than $17,500 per violation. Violations are defined as improperly disclosing an individual's personal information without consent or court order. With Kuwait and the entire Arabian region trying to shift their economies away from oil and towards FinTech and other modern enterprises, data is now the most valuable resource in Kuwait. Accordingly, the country finally passed comprehensive data protection regulations in 2021.

5.2.11 Malaysia

Malaysia's first comprehensive data privacy legislation came into effect in 2013. The *Personal Data Protection Act 2010* consists of seven key points that protect personal and private data. These are the General Principle, the Security Principle, the Disclosure Principle, the Access Principle, the Notice and Choice Principle, the Retention Principle and the Data Integrity Principle. Under the regulation, for consent to be valid, the subject must receive information about their rights, written notice for the data collection and details about who will access their data. A definite change between Malaysia's regulation and the GDPR is that there is no requirement for companies to appoint a data protection officer.

In April 2017, a *Public Consultation Paper* (PCP) was issued on *Personal Data Protection (Transfer of Personal Data to Places Outside Malaysia)*. The draft lists several countries considered adequate for the transfer of personal data, including European countries, countries considered adequate by the EU, numerous local countries (e.g. Australia, Japan) and the United States. Key nearby markets that are not on the list include India and China. In February 2020 the Malaysian released a proposal to amend the PDPA which would simplify data transfer.

Malaysia has a de facto data localisation policy because the government never released a final list of acceptable countries for cross-border data transfer. Personal information regarding Malaysian citizens must be stored locally.

5.2.12 Russia

Russia was one of the first countries to pass federal data protection laws, passing *Federal Law No. 152-FZ on Personal Data* on 27 July 2006.[19] Since then, the law has been amended and other data laws, such as 242-FZ and 149-FZ, have also been passed. With even the 1993 Constitution mentioning data privacy (Articles 23 and 24), the matter has long been important within Russia. Russia has spent much of 2021 trying to enforce 152-FZ on tech giants like Google, Facebook and Twitter, handing out fines and warnings left and right. While Roskomnadzor has been focused on these select companies recently, their scope may expand to other countries if the government continues prioritising data security.

The information is considered personal data if it identifies a specific person. The rules of localisation apply to companies only if they carry out specific actions intentionally: collecting, recording, systematisation, accumulating, storing, clarifying (updating and modifying) and extracting personal data. Yet, Russian data residency law does not prohibit further processing of Russians' personal data abroad if this data was previously included in a Russian database and is updated there as necessary. Therefore, personal data usage, transfer (distribution, provision),

[19] https://pd.rkn.gov.ru/authority/p146/p164/

depersonalisation, blocking, removal or destruction can be performed using databases outside of Russia. As one of the regulated industries in Russia, payment data regulations specify that all credit cards including international ones and all domestic transactions are handled via the Russian National Payment System.[20] As well, internal transactional records should be processed and stored within the Russian Federation.

Russia's data residency requires companies to store and process the data of Russian citizens locally, but if that mandate is fulfilled, they are permitted to transfer data out of Russia, provided it goes to a country with adequate data regulations. This list conspicuously excludes the United States, so companies cannot transfer data from Russia to America. Technically, data localisation law applies to all Russian companies, branches and representative offices of foreign corporations, as well as other legal entities incorporated outside Russia, that do not have an official presence in Russia but have business activities in the local market. This law also applies if an international company uses the domain names ".ru and. рф", has a Russian-language website, receives payment in Russian rubles or delivers goods to the Russian Federation. So, any company doing business in Russia or with Russians may be affected by the law, even if it is not registered in Russia.

Roskomnadzor is the official governmental body that controls how companies fulfil their localisation duties in Russia. Among its functions is the ability to launch an off- or on-site audit. During the audit, a state department usually examines notifications sent by the operator and can require any necessary information—for example, confirmation of the place where databases are stored. While Roskomnadzor is the presiding authority over Russia's various data protection laws, 161-FZ specifically regulates payment card data and is thus regulated by the Central Bank of Russia. The process of obtaining client consent is laid out in 152-FZ, the country's original data protection law. Transparent consent is required, and individuals hold the right at any time to request a company delete their personal data.

Data protection rules for Russia are found in specific legislation, particularly the Data Protection Act No. 152-FZ dated 27 July 2006 (DPA) as well as various regulatory acts adopted to implement the DPA. Some other data residency laws for Russia are the *Information, Information Technologies and Information Protection Act No. 149-FZ* dated 27 July 2006, which establishes basic rules as to the information in general and its protection.

Compliance with *Data Protection Act No. 152* requires involvement of lawyers as well as IT specialists in order to create systems and policies for dealing with personal data. Companies which already carry out activities in Russia need to review their database content and systems for dealing with personal data. Some databases obviously need to be relocated to Russia, and moving them to the cloud will not solve the issue based on local practice. The financial penalties from breaching any of Russia's data protection laws are not crippling, with each offence usually carrying a

[20] https://nspk.com/

fine of $1500–$8000. However, the government can issue multiple fines over a short span, creating a host of if not financial, then bureaucratic issues for a company. Any company that does not directly comply will be targeted until it submits, as we have seen recently with Apple and Google.

5.2.13 Saudi Arabia

The primary source of law in the Kingdom of Saudi Arabia (KSA) is Shari'ah principles. These are Islamic principles derived from the Holy Quran and the Sunnah. Besides Shari'ah principles, the law in the KSA consists of secular regulations passed by the government.

Generally, sensitive customer content or content from private sector-related industries cannot be transferred outside the country. Cross-border data transfers might be regulated differently according to industry-specific rules, however. For example, all personal health information must only be stored in Saudi Arabia. Banks are required to store certain records in the country as well. Other personal identifiable information may only be processed or transferred outside the country after obtaining written approval in coordination with the Saudi Data and AI Authority (NDMO).[21] It is unclear if every cross-border transfer will need written approval, which would essentially kill such transfers. It is for this reason that Saudi data regulations should be considered strict until otherwise noted.

The new *Personal Data Protection Law* requires users to consent to the collection of their data and gives them significant rights. Data subjects can withdraw consent at any time, request access to their data on record and must be informed if their data will be stored, processed or transferred outside the country. The law also calls for data minimisation, further anti-spam measures and the need for data controllers to destroy personal data after the collection purpose has been achieved.

The KSA *Cloud Computing Regulatory Framework* (CCF) is based on international best practices and governs the rights and obligations of cloud service providers (CSPs), individual customers, government entities and businesses. The CCF is one of only a few examples of cloud-specific regulatory frameworks around the world and includes principles of data protection. Some of the provisions, such as security breach notification, are in line with the approach taken in the EU, while others, such as the requirement to register with the Communications and Information Technology Commission (CITC) content classification, are specific to KSA.

Some of the most important features of the CCF from a data protection perspective are the cloud security requirements cloud service providers must adhere to. Cloud customer information can be subject to different levels of information security, depending on the required level of preservation of the information's confidentiality, integrity and availability. CSPs must also inform any cloud

[21] https://sdaia.gov.sa/

customer, upon request, of the information security features offered by the CSP or applied to the cloud customer's information.

There are certain secular regulations passed by the government, which, although not dedicated as a whole to data privacy/protection, contain specific provisions governing the right to privacy and data protection in certain contexts. There may also be specific regulations applicable to certain industries, for example, in banking, which is regulated by the Saudi Arabian Monetary Authority (SAMA).[22] More specific examples of privacy-related legal provisions are found in the *Anti-Cyber Crime Law of 2007* (Royal Decree No. M/17) and the new *E-commerce Law of 2019.*[23] In addition, sectoral regulations contain data protection obligations regarding organisations working in telecommunication, IT/cloud services, healthcare and financial services industries.

Financially, Saudi Arabia's *Rules on Outsourcing*,[24] which is regulated by the Saudi Central Bank (SAMA),[25] states that banks and institutions handling payment information need to process the data themselves and cannot transfer data without authorisation from a regulator.

Saudi Arabia finally has an overarching data protection law, with the bill passing on 14 September 2021. However, the bill's implementation was recently delayed, and will only become effective 17 March 2023, with the year-long delay designed to allow more time for necessary public discourse on the matter of data protection. While we now need to wait and see how Saudi Arabia handles the law once in effect, passing a comprehensive law with expanded individual rights is sure to make the regulations somewhat stricter and/or more intensive for companies to keep up with. Eventual enforcement of the bill will fall on the Saudi Data and AI Authority (NDMO).

Having already passed its data protection law, and thanks to tighter data regulations, among other things, the Kingdom of Saudi Arabia believes it is on the way towards maximising its potential as a global player in the technological sphere.

5.2.14 Switzerland

The main legislation in Switzerland on data regulation is the *Swiss Data Protection Act,* known as revDPA. First passed in 1993, updates were officially passed in Autumn 2020 and will take effect sometime in 2022. While not an exact copy, the

[22] http://www.sama.gov.sa/en-US/Pages/default.aspx

[23] https://platform.dataguidance.com/legal-research/anti-cyber-crime-law-2007-royal-decree-no-m17

[24] https://www.sama.gov.sa/en-US/Laws/FinanceRules/Outsourcing%20Rules%20-%20Revised%20v2%20Final%20Draft-Dec-2019.pdf

[25] http://sama.gov.sa/en-us/pages/default.asp

majority of the law's framework resembles the EU's *General Data Protection Regulation*.[26]

Similar to the *General Data Protection Regulation*, companies must keep records of data processing, must report any data breach and must conduct data protection impact assessments. However, companies do not have to appoint data protection officers, and data transfers abroad will require less oversight than the EU demands.

Cross-border data transfers are allowed to countries with adequate data protection. If a country lacks proper legislation, such as the United States, a special agreement (SCC) needs to be put in place before data can travel abroad to that location. This is perhaps most strict regarding customer banking information, as the regulator, the Swiss Financial Market Supervisory Authority (FINMA), has placed a renewed emphasis on protecting customer data.

Similar to the GDPR, client consent lies at the heart of revDPA. Companies must lawfully acquire consent before processing customer data, and customers have the right at any time to revoke that right or request to see their information on record.

Perhaps the most unique thing about Swiss data protection penalties is that they are levied against individuals and not companies. The maximum fine per violation is roughly $270,000 (the maximum fine was raised significantly in the 2020 revisions). FDPIC does not have the authority to levy fines however, as that power rests with Cantonal law enforcement, which makes fines a rare occurrence.

The Schrems II ruling in 2020 ended not only the EU-US *Privacy Shield*, but the Swiss-US one as well. Despite revDPA's recent revisions, companies are still waiting on clarity regarding how standard contractual clauses (SCCs) will work, although the Swiss government has announced SCCs will still be accepted. Any further guidelines will greatly influence how companies conduct future Swiss American business transactions.

5.2.15 Turkey

The main data protection law in Turkey is *Law No. 6698*, enforced by the Personal Data Protection Authority (KVKK). Law No. 6698 includes any information related to an identified or identifiable individual.

KVKK applies to any data controllers and data processors that collect data or process data collected from Turkey. Any company handling such information must comply with Law No. 6698, regardless of industry. Additional industry-related data protection laws exist, but do not supersede Law No. 6698.

[26] *EU General Data Protection Regulation (GDPR):* Regulation (EU) 2016/679 of the European Parliament and of the Council of 27 April 2016 on the protection of natural persons with regard to the processing of personal data and on the free movement of such data, and repealing Directive 95/46/EC (General Data Protection Regulation), OJ 2016 L 119/1.

Turkey's 2020 Internet bill required social media companies to set up a legal entity within the country to ensure Turkish users' data stays within the country. While some, like Twitter, have complied with the law, other companies have not been fast enough in achieving compliance with Turkey's many data regulations. Such was the case in September 2021, when KVKK fined WhatsApp over $200,000 for violating Law No. 6698, the country's main data regulations.

Cross-border data transfers require the user to give explicit consent and that the data be sent to a country with adequate levels of data protection as determined by the Turkey Data Protection Authority (TDPA). This corresponds to a high level of data residency requirements. Certain industries are affected more than others by the wide scope with which Turkey governs data, with banking, telecommunications and healthcare under particular focus due to the need to comply with multiple regulations.

The Banking Regulation and Supervision Agency (BRSA) also enforces Law No. 6493, which notes companies providing e-payments must have a local base to keep the information systems used for the activities within Turkey. Regulations imposed by the KVKK carry fines of hundreds of thousands of dollars and even prison time in more serious cases, if companies fail to comply.

Turkey's 2020 internet bill required social media companies to set up a legal entity within the country to ensure Turkish users' data stays within the country. While some, like Twitter, have complied with the law, other companies have not been fast enough in achieving compliance with Turkey's many data regulations. Such was the case in September 2021, when KVKK fined Whatsapp over $200,000 for violating law no. 6698, the country's main data regulations.

5.2.16 United Arab Emirates (UAE)

The UAE Government is currently developing its new *Federal Data Protection Law*. This is a unique opportunity for the country to build a framework for data privacy law that fosters an innovative business environment.

UAE has some specific approaches to managing data localisation laws. First of all, the UAE does not have a comprehensive data protection law at the federal level, yet there are a number of laws in place that govern privacy and data security in this country. There are also sector-specific data protection provisions in certain laws. The UAE has several special economic or sector free zones, three of which have specific data protection laws. These are the Dubai International Financial Centre, the Abu Dhabi Global Market (ADGM) and the Dubai HealthCare City.

Data protection in UAE is governed by federal laws and regulations from UAE Central Bank and Telecommunications Regulatory Authority (TRA). These UAE federal laws and regulations contain various provisions in relation to privacy and the protection of personal data. Some of them are as follows:

- The Cyber Crime Law—Federal Decree Law No. 5 of 2012. The Cyber Crime Law criminalises obtaining, possessing, modifying, destroying or disclosing (without authorisation) electronic documents or electronic information relating to medical records (Article 7).
- Penal Code (Federal Law No. 3 of 1987 as amended).
- UAE's Central Bank Regulatory Framework for Stored Values and Electronic Payment Systems ("Digital Payment Regulation") 1 January 2017.
- Telecommunications Regulatory Authority (TRA)—The Consumer Protection Regulations, Version 1.3, 10 January 2017.
- The DHCC Health Data Protection Regulation No. 7 of 2013.
- The DIFC implemented DIFC Law No. 1 of 2007 Data Protection Law in 2007 which was subsequently amended by DIFC Law No. 5 of 2012 Data Protection Law Amendment Law ("DPL").
- The Dubai Data Law, 27 December 2015.

The UAE data localisation policy is based on assumptions that localising the data makes it safer and prevents foreign law enforcement from accessing data. It is also seen as a means of attracting and locking ICT investment into the country.

5.2.17 United States

In the United States, there is no comprehensive federal data protection law; however there are sectoral laws relating to data residency/localisation or government access to certain types of data. In addition, many of the obstacle's organisations faced regarding the use of American citizens' data abroad and data transfers to the United States were addressed by mechanisms, including the EU-US Privacy Shield. However, because of the Court of Justice of the European Union ("CJEU") decision in *Data Protection Commissioner v. Facebook Ireland Limited*, Maximillian Schrems (C-311/18) ("the Schrems II case"), the EU-US Privacy Shield was invalidated, leaving a significant regulatory gap for data transfers to and/from the United States.

Examples of explicit laws in the United States include the US *Cloud Act* and *Patriot Act*. Per the *Cloud Act*, all data in the possession, custody or control, pursuant to lawful process—regardless of the location of the data—is subject to access for enforcement purposes using bilateral agreements entered with the US Department of Justice and regulatory counterparts. Similarly, the US *Patriot Act* established that most categories of personal information, including phone records, computer records, credit history and banking history, are subject to access by US government agencies for the prevention and mitigation of terrorism.

5.2.18 *Uzbekistan*

The former Soviet Republic of Uzbekistan, home to nearly 35 million people and an emerging market in Central Asia, just introduced data localisation requirements. The *Law of the Republic of Uzbekistan On Personal Data*, dated 2 July 2019, No. O'RQ-547 went into effect in April 2021. The new law, as of now, is enforced by the State Personalization Center under the Cabinet of Ministers of the Republic of Uzbekistan.

The *Law On Personal Data* mandates the local processing and storage of personal data "on technical means physically located on the territory of Uzbekistan" as well as on databases registered with the data protection authority in Uzbekistan. However, despite those restrictions, mirrored copies of data may be sent abroad, so cross-border data transfers are allowed. Additional previous laws, such as the *Law on E-Commerce* and the *Law on Electronic Document Management*, also require records to be stored and not transferred to third parties, though neither law introduced strict data residency requirements until the comprehensive law went into effect in 2021.

Client consent is required and can be given in multiple forms. Additionally, individuals can withdraw their consent and be ensured their data is destroyed, as well as request information on the grounds and purpose of data processing, the storage period of data and any intended cross-border transfers. If a customer agrees to a cross-border transfer, any security requirement for the receiving end is waived.

Administrative penalties carry initial fines of between $125 and 250. Repeat offences are treated much more harshly and migrate to the criminal code instead of the administrative one. These fines are multiplied by $10 to $1250 and could carry correctional work or deprivation of certain rights (such as to conduct business that handles personal data).

Central Asia is a complicated place to do business internationally, and places like Uzbekistan, which had a web of intersecting laws but no overarching one, needed to clarify its data governance requirements. Despite low fines and some clarification about its data protections, Uzbekistan is still a challenging market to enter and will need further legislation to sort out where international companies fit in its data governance scheme.

5.2.19 *Vietnam*

Vietnam does not have a comprehensive data protection law at the moment, but the government released the draft decree of such a bill in February and hopes to have it signed into law sometime in 2022. In the meantime, there has been lax enforcement of existing legislation as the country awaits the new, much tougher regulations to take effect.

Data localisation laws in Vietnam have certain specifics. Unlike other countries, there is no single comprehensive data protection law in this country. Instead, data protection regulations are currently spread across various acts and guiding legal documents. Among them are the following: the *Civil Code* (2015), the *E-Transactions Law* (2005), the *Information Technology Law* (2006), the *Consumer Rights Protection Law* (2010), the *Cyber Information Security Law* (2015) and the *Cybersecurity Law* (2018). In these circumstances, it is sometimes difficult for companies to identify if data protection laws are applicable to each situation.

The key principles on collection, storage, use, processing, disclosure or transfer of personal information are specified in the following main laws and guiding documents:

- Law No. 24/2018/QH14 on Cybersecurity, passed by the National Assembly on 12 June 2018 (*Cybersecurity Law*)
- Law No. 86/2015/QH13 on Cyber Information Security, passed by the National Assembly on 19 November 2015; amended by Law No. 35/2018/QH14 dated 20 November 2018, on amendments to some articles concerning planning of 37 Laws (*Cyber Information Security Law*)
- Decree No. 85/2016/ND-CP dated 1 July 2016, on the security of information systems by classification
- Circular No. 20/2017/TT-BTTTT dated 12 September 2017 of the Ministry of Information and Communications, providing regulations on coordinating and responding to information security incidents nationwide

Each industry in Vietnam may also have their own respective regulating documents. In other words, applicability of legal documents will depend on the actual context of each business case, e.g. profile, health, finance or employee data can be subject to specialised data protection regulations depending on the industry. There's also special regulation on employees' personal information as stated in *Labour Code 2012*.

The most important Vietnamese legal document that regulates data protection is the *Cybersecurity Law of Vietnam.* Unlike other cybersecurity laws that were inspired by the GDPR of the EU, this law shares similarities with China's *Cybersecurity Law* established in 2017.[27] It focuses on providing the government with the ability to control the flow of information, instead of enforcing data privacy rights for private data subjects. Other laws that are currently being prepared are a draft decree detailing a number of articles of the *Cybersecurity Law*, a draft Decree detailing the order of and procedures for application of a number of cybersecurity assurance

[27] https://incountry.com/blog/data-protection-laws-in-china-ways-to-stay-compliant-with-cloud-data/

measures and a draft decision of the Prime Minister promulgating a list of information systems important for national security.

As of now, the data of Vietnamese citizens needs to stay in Vietnam, but data can be copied and transferred elsewhere. The *Decree on Personal Data Protection* will trump all existing legislation and splits data into basic personal and sensitive personal, the latter including typical PII fields like gender, religion, biometrics, etc. The Decree will require companies to cooperate with the Personal Data Protection Commission, an entity formed by the Ministry of Public Security (MPS) to enforce the *Decree.*

Client consent is mandatory for companies to receive before sending data abroad. The new draft decree has even specified forms of consent, a fact that will likely grant Vietnamese citizens more power to limit data collection. Current penalties include fines from $1000 to $50,000 and a 3–7-year prison sentence. The monetary figures are expected to rise considerably per violation once the draft decree becomes law, with fines of up to 5% of earned revenue included therein. With China, India and other neighbouring Asian countries also passing stricter regulations, Vietnam following suit is a sign that the global baseline for data regulations is changing rapidly.

5.3 Data Residency Comparative

To help explain the different methods of dealing with data residency and localisation, we can look at some of the strategies used by a select group of countries to address this issue. A comparative of data residency provisions is provided in Table 5.1, based on the information gathered and presented with regard to the countries in the previous sections. In this table, we compare the practices of different countries based on severity, regulated industries, data residency requirements, consent and affiliated penalties.

While the definition of privacy varies across regions, there is one thing everyone agrees on—that privacy is important. While it is always challenging to see into the future, the last 5 years have been indicative of the likelihood of a continued increase in regional differences in privacy laws in the years to come. There is a clear trend in regulating specific industries, some more than others, with profile and health data being among the most commonly protected. In general, we can see a pattern emerging of countries taking a protectionist approach to cross-border data transfers, which is understandable given the rapidly changing nature of our technical and digital world.

Table 5.1 Data residency comparative

Country	Regulated industries	Residency requirements	Client consent for data	Penalties for noncompliance
China severity: high	• Profile. • Health • Employee • Finance • Payment	Cross-border data flows are largely not permitted	Companies must get client consent to store and process data, even if the data will remain within China's borders	Fines for noncompliance are massive: up to $7.7 million or 5% of a company's previous year revenues
Germany severity: high	• Profile • Health • Employee • Finance • Payment.	Cross-border data transfers are permitted under the GDPR, as long as the data is going to a country with adequate data protection	Consent must be freely given, informed and unambiguous and state the specific reasons for data processing	While Germany has been one of the leading forces behind European data protections, the country has yet to institute and enact severe penalties
India severity: medium	• Profile • Health • Employee • Finance • Payment.	While there are currently no restrictions banning cross-border data transfers, payment card data and sensitive personal data must be stored domestically	Cross-border data transfers require explicit user consent to occur in addition to the Data Protection Authority's role in approving transfers abroad	Failure to conduct a proper data audit is punishable with a fine of over $650,000 or 2% of annual fiduciary turnover (whichever is higher), in addition to potential operating shutdowns for noncompliance
Indonesia severity: medium	• Profile • Health • Employee • Finance • Payment	In certain cases, cross-border data transfer is allowed Healthcare data processing can also be done abroad, as long as it is domestically accessible, although it must be stored domestically	The sector-specific regulations currently in place do not handle the matter of client consent, but upcoming legislation is anticipated to have a heavy focus on offering citizens GDPR-style rights, including client consent	Without an overarching data protection bill, penalties for noncompliance don't have a numerical figure on them yet. Violations of sector-specific laws will result in warnings and, eventually, sanctions against the offending organisations
Japan severity: medium	• Profile • Health • Employee • Finance • Payment	There are no registration requirements for the overseas transfer of data and no general data localisation	Consent must be freely given and informed and state the specific reason, especially for cross-border transfers	Penalties can include both criminal and civil actions

(continued)

Table 5.1 (continued)

Country	Regulated industries	Residency requirements	Client consent for data	Penalties for noncompliance
		requirements in Japanese laws		
Kuwait severity: medium	• Profile • Health • Employee • Finance	While data protections are not unbending, there are certain types of data, defined by the Data Classification Policy, which cannot be transferred abroad	Companies must receive written client consent before transferring data abroad, in addition to informing individuals why their data is being transferred and who is receiving it	There have yet to be publicised occurrences of any violation of the new laws, but regarding Law No. 20 of 2014, penalties are a maximum of 3 years in prison and a fine of no less than $17,500 per violation
Russia severity: high	• Profile • Health • Employee • Finance • Payment	Russia's data residency requires companies to store and process the data of Russian citizens locally. They are then permitted to transfer data out of Russia, provided that it goes to a country with adequate data regulations	Transparent consent is required, and individuals hold the right at any time to request a company delete their personal data	The financial penalties from breaching any of Russia's data protection laws are not crippling, with each offence usually carrying a fine of $1500–$8000. The government can also issue multiple fines over a short span
Saudi Arabia severity: high	• Profile • Health • Employee • Finance • Payment	In general, sensitive customer content or content from private sector-related industries cannot be transferred outside the country. Cross-border transfers may be regulated differently by industry-specific rules and after obtaining approval with the Saudi Data and AI Authority (NDMO)	The new *Personal Data Protection Law* requires users to consent to the collection of their data and gives them significant rights	Violations of the provisions relating to the transfer or disclosure of data potentially carry a 1-year prison sentence and/or a fine of up to SAR one million. Unlawful disclosure carries a 2-year prison sentence and/or fine of up to SAR three million. Any violation of the law aside from these specific occurrences can carry a fine up to SAR five million
	• Profile • Health	Cross-border data transfers are allowed	Companies must lawfully acquire	Perhaps the most unique thing about

(continued)

Table 5.1 (continued)

Country	Regulated industries	Residency requirements	Client consent for data	Penalties for noncompliance
Switzerland severity: medium	• Employee • Finance	to countries with adequate data protection. If a country lacks proper legislation, a special agreement needs to be put in place before data can travel to that location	consent before processing customer data, and customers have the right at any time to revoke that right or request to see their information on record	Swiss data protection penalties is that they are levied against individuals and not companies. The maximum fine per violation is roughly $270,000
Turkey severity: high	• Profile • Health • Employee • Finance • Payments	Cross-border data transfers require the user to give explicit consent and that the data be sent to a country with adequate levels of data protection, determined by the Data Protection Authority	Explicit consent, a crucial part of Law No. 6698, must be acquired through a user's free will	Regulations imposed by the KVKK carry fines of hundreds of thousands of dollars and even prison time in more serious cases, if companies fail to comply
United Arab Emirates (UAE)	• Profile • Health • Finance • IoT	The UAE data localisation policy is based on assumptions that localising the data makes it safer and prevents foreign law enforcement from accessing data	The United Arab Emirates (UAE) has no overall data protection law; however its civil code does require the user consent to data transfer	Violations of the healthcare sector law will result in civil penalties
Uzbekistan severity: medium	• Profile • Health • Employee • Finance	Mirrored copies of data may be sent abroad, so cross-border data transfers are allowed	Client consent is required and can be given in multiple forms	Administrative penalties carry initial fines of between $125 and $250. Repeat offences are treated more harshly and migrate to the criminal code. These fines are multiplied by $10 to $1250 and could carry correctional work or deprivation of certain rights
Vietnam severity: high	• Profile • Health • Employee	As of now, the data of Vietnamese citizens needs to stay in Vietnam, but data can be copied and	Client consent is mandatory for companies to receive before sending data abroad. The new	Current penalties include fines from $1000 to $50,000 and a 3–7-year prison sentence.

(continued)

Table 5.1 (continued)

Country	Regulated industries	Residency requirements	Client consent for data	Penalties for noncompliance
	• Finance • Payment	transferred elsewhere	draft decree has even specified forms of consent, a fact that will likely grant Vietnamese citizens more power to limit data collection	Noncompliance in terms of data protection may be subject to sanctions (even criminal ones), depending on the severity

5.4 Summary

In this chapter we discussed some of the major technical challenges in data sovereignty and digital governance. We explored the current challenges in domestic law, as well as the challenges encountered in international law and global Internet standards. Beyond merely legal perspectives, we also considered the ethical, moral and rights-based arguments relating to access to information, freedom of expression and digital censorship within our international legal landscape.

In the next chapter, we will explore the concept and realisation of digital governance and its intersection with data sovereignty on both a national and an international level.

The answers to the following questions are provided within this chapter:

1. What are the most commonly regulated industries?
2. Which industry is the MOST commonly regulated?
3. Why do you think that is?
4. What are some examples of the differing strategies between nations in addressing consent?
5. How does the GDPR method differ from other methods?

Chapter 6
Emerging Topics in Data Sovereignty and Digital Governance

In an increasingly digital world, we are continuously being met with novel and unprecedented challenges, the likes of which we have never encountered in our human history. Not only are we engaged in the balancing act of protecting individual and personal rights and freedoms with respect to both positive and negative rights, but also we are working to balance the interests of our communities, societies, and the greater global world in this novel forum. We have seen a growth in the desire for equality and equity for under-recognized and marginalised social groups, including the interest in indigenous rights and data sovereignty, as well as greater environmental accountability and transparency regarding the inner workings of data centres and large technological infrastructure projects. Finally, we are seeing the fast-paced growth of artificial intelligence, machine learning, and the corresponding concerns relating to human rights and ethics. These topics, and many more, are the ones which we will have to face as emerging topics in data sovereignty and digital governance in the months and years to come.

6.1 Topic Overview

In this chapter, we will discuss some of the major emerging topics in data sovereignty and digital governance, including the concept of digital rights, indigenous data sovereignty, the issues associated with data centres and data mines, digital self-determination, artificial intelligence, social responsibility, and data policing.

M. Lukings, A. Habibi Lashkari, *Understanding Cybersecurity Law in Data Sovereignty and Digital Governance*, Progress in IS,
https://doi.org/10.1007/978-3-031-14264-2_6

6.2 Digital Rights

We can classify our rights as individuals into two major groups: moral/natural/human rights, and legal/procedural rights. Moral rights are based on the idea of rights which are "natural" or inherent to all humans, also called human rights. Legal rights refer to our rights and obligations within the law and legal systems.

Digital rights are those human and legal rights that allow individuals to access, use, create, and publish digital media or to access and use computers, other electronic devices, and telecommunications networks. The concept is particularly related to the protection and realisation of existing rights, such as the right to privacy and freedom of expression, in the context of digital technologies, especially the Internet.[1] The laws of several countries recognize a right to Internet access.[2]

A number of human rights have been identified as relevant with regard to the Internet. These include freedom of expression, privacy, and freedom of association.[3] Furthermore, the rights to education and multilingualism, consumer rights, and capacity building in the context of the right to development have also been identified.[4]

The *APC Internet Rights Charter* was established by the Association for Progressive Communications (APC) at the APC Europe Internet Rights Workshop, held in Prague, February 2001. The *Charter* draws on the *People's Communications Charter* and develops seven themes: Internet access for all; freedom of expression and association; access to knowledge, shared learning, and creation—free and open-source software and technology development; privacy, surveillance, and encryption; governance of the Internet; awareness, protection, and realisation of rights.[5,6]

The APC states that "the ability to share information and communicate freely using the Internet is vital to the realization of human rights as enshrined in the *Universal Declaration of Human Rights*, the *International Covenant on Economic, Social, and Cultural Rights*, the *International Covenant on Civil and Political Rights* and the *Convention on the Elimination of All Forms of Discrimination against Women*." The *APC Internet Rights Charter* is an early example of a so-called Internet bill of rights, an important element of digital constitutionalism.

[1] "Digital freedom: the case for civil liberties on the Net". *BBC News*. 1999-03-04.

[2] Lucchi, Nicola (2011-02-06). "Access to Network Services and Protection of Constitutional Rights: Recognizing the Essential Role of Internet Access for the Freedom of Expression". Rochester, NY.

[3] Benedek 2008, 17 November 2011.

[4] Kim, Minjeong; Choi, Dongyeon (2018). "Development of Youth Digital Citizenship Scale and Implication for Educational Setting". *Journal of Educational Technology & Society*. 21 (1): 155–171.

[5] "Towards a charter for Internet rights". Internet Rights UK.

[6] Benedek, Wolfgang; Veronika Bauer; Matthias Kettemann (2008). *Internet Governance and the Information Society*. Eleven International Publishing. p. 39.

In December 2003, the World Summit on the Information Society (WSIS) was convened under the auspices of the United Nations (UN). After lengthy negotiations between governments, businesses, and civil society representatives, the WSIS *Declaration of Principles* was adopted, reaffirming human rights[7]:

> We reaffirm the universality, indivisibility, interdependence and interrelation of all human rights and fundamental freedoms, including the right to development, as enshrined in the Vienna Declaration. We also reaffirm that democracy, sustainable development, and respect for human rights and fundamental freedoms as well as good governance at all levels are interdependent and mutually reinforcing. We further resolve to strengthen the rule of law in international as in national affairs.

The WSIS *Declaration* also makes specific reference to the importance of the right to freedom of expression in the "Information Society" by stating[8]:

> We reaffirm, as an essential foundation of the Information Society, and as outlined in Article 19 of the Universal Declaration of Human Rights, that everyone has the right to freedom of opinion and expression; that this right includes freedom to hold opinions without interference and to seek, receive, and impart information and ideas through any media and regardless of frontiers. Communication is a fundamental social process, a basic human need and the foundation of all social organisation. It is central to the Information Society. Everyone, everywhere should have the opportunity to participate and no one should be excluded from the benefits the Information Society offers.

The 2004 WSIS *Declaration of Principles* also acknowledged that "it is necessary to prevent the use of information resources and technologies for criminal and terrorist purposes, while respecting human rights".[9] Wolfgang Benedek comments that the WSIS *Declaration* only contains a number of references to human rights and does not spell out any procedures or mechanisms to assure that human rights are considered in practice.[10]

In 2005, the United Kingdom's Open Rights Group published a digital rights landscape, documenting the range of organisations and people active in the cause of preserving digital rights.[11]

6.2.1 Moral/Human Rights

Moral rights, or **human rights**, are the basic rights and freedoms that belong to every person in the world, from birth until death. They apply regardless of where you are from, what you believe or how you choose to live your life. They can never be

[7] Klang, Mathias; Murray, Andrew (2005). *Human Rights in the Digital Age*. Routledge. p. 1.

[8] Klang, Mathias; Murray, Andrew (2005). *Human Rights in the Digital Age*. Routledge. p. 1.

[9] Klang, Mathias; Murray, Andrew (2005). *Human Rights in the Digital Age*. Routledge. p. 2.

[10] Benedek, Wolfgang; Veronika Bauer; Matthias Kettemann (2008). *Internet Governance and the Information Society*. Eleven International Publishing. p. 36.

[11] Charman-Anderson, Suw (2005-09-17), *UK digital rights landscape*.

taken away, although they can sometimes be restricted—for example, if a person breaks the law, or in the interests of national security. These basic rights are based on shared values like dignity, fairness, equality, respect, and independence, which are defined and protected by law.

Human rights are based on moral principles or norms for certain standards of human behaviour and are regularly protected in municipal and international law. Human rights are moral principles or norms for certain standards of human behaviour and are regularly protected in municipal and international law. They are commonly understood as inalienable, fundamental rights "to which a person is inherently entitled simply because she or he is a human being" and which are "inherent in all human beings", regardless of their age, ethnic origin, location, language, religion, ethnicity, or any other status. They are applicable everywhere and at every time in the sense of being universal, and they are egalitarian in the sense of being the same for everyone. They are regarded as requiring empathy and the rule of law and imposing an obligation on persons to respect the human rights of others, and it is generally considered that they should not be taken away except as a result of due process based on specific circumstances.

The true forerunner of human rights discourse was the concept of natural rights which appeared as part of the mediaeval natural law tradition that became prominent during the European Enlightenment. From this foundation, modern human rights arguments emerged over the latter half of the twentieth century.

The *Magna Carta* is an English charter originally issued in 1215 which influenced the development of the common law and many later constitutional documents related to human rights, such as the *1689 English Bill of Rights*, the *1789 United States Constitution*, and the *1791 United States Bill of Rights*.

The seventeenth-century English philosopher John Locke discussed natural rights in his work, identifying them as being "life, liberty, and estate (property)", and argued that such fundamental rights could not be surrendered in the social contract. In Britain in 1689, the *English Bill of Rights* and the *Scottish Claim of Right* each made a range of oppressive governmental actions illegal. Two major revolutions occurred during the eighteenth century, in the United States (1776) and in France (1789), leading to the *United States Declaration of Independence* and the *French Declaration of the Rights of Man and of the Citizen*, respectively, both of which articulated certain human rights. Additionally, the V*irginia Declaration of Rights of 1776* encoded into law a number of fundamental civil rights and civil freedoms.

> We hold these truths to be self-evident, that all men are created equal, that they are endowed by their Creator with certain unalienable Rights, that among these are Life, Liberty and the pursuit of Happiness.—United States Declaration of Independence, 1776.

The *Universal Declaration of Human Rights* (UDHR) is a non-binding declaration adopted by the United Nations General Assembly in 1948, partly in response to the events of World War II.[12] The UDHR urges member states to promote a number

[12](A/RES/217, 10 December 1948 at Palais de Chaillot, Paris).

of human, civil, economic, and social rights, asserting that these rights are part of the "foundation of freedom, justice, and peace in the world". The declaration was the first international legal effort to limit the behaviour of states and make sure they did their duties to their citizens following the model of the rights-duty duality.

> . . .recognition of the inherent dignity and of the equal and inalienable rights of all members of the human family is the foundation of freedom, justice, and peace in the world.—Preamble to the Universal Declaration of Human Rights, 1948.

In Canada, along with the constitutional provisions of the *Charter of Rights and Freedoms,* there is also the *Canadian Human Rights Act.*[13] The *Canadian Human Rights Act* was first created in 1977 to ensure equality of opportunity and prohibit discrimination on the basis of race, age, sex, and a variety of other categories—called "protected grounds".[14] Unlike the *Canadian Charter of Rights and Freedoms*, which provides Canadians with a wide range of positive and negative rights, the *Canadian Human Rights Act* specifically covers equality rights and only governs that which falls under federal jurisdiction. With regard to the provinces, each province and territory in Canada also has its own human rights legislation, which applies to provincial jurisdictions, including activities in hospitals and in schools. The *Canadian Human Rights Act* has also been used to create regulations for: age guidelines; customs and excise human rights investigations; equal wages; human rights benefits; immigration investigations; and rules of procedure for human rights tribunals.

The purpose of the *Canadian Human Rights Act* is to ensure that all individuals have "opportunity equal with other individuals to make for themselves the lives that they are able and wish to have and to have their needs accommodated, consistent with their duties and obligations as members of society, without being hindered in or prevented from doing so by discriminatory practices."[15] The *Act* sets out 13 prohibited grounds of discrimination: "race, national or ethnic origin, colour, religion, age, sex, sexual orientation, gender identity or expression, marital status, family status, genetic characteristics, disability, and conviction for an offence for which a pardon has been granted or in respect of which a record suspension has been ordered."

6.2.1.1 Right to Privacy

The **right to privacy** is an element of various legal traditions that intend to restrain governmental and private actions that threaten the privacy of individuals. Over 150 national constitutions mention the right to privacy.[16]

---

[13] *Canadian Human Rights Act*, second Session, 39th Parliament, 2007.

[14] *Canadian Human Rights Act* (RSC, 1985, c H-6) Available at: https://laws-lois.justice.gc.ca/eng/acts/h-6/

[15] *Canadian Human Rights Act* (RSC, 1985, c H-6) Available at: https://laws-lois.justice.gc.ca/eng/acts/h-6/

[16] "Read about "Right to privacy" on Constitute". *constituteproject.org*

The concept of a human "right to privacy" began when the Latin word "ius" expanded from meaning "what is fair" to include "a right—an entitlement a person possesses to control or claim something," by the Decretum Gratiani in Bologna, Italy in the twelfth century.[17]

In the United States, an article in the 15 December 1890 issue of the *Harvard Law Review*, written by attorney Samuel D. Warren and future U.S. Supreme Court Justice, Louis Brandeis, entitled "The Right to Privacy", is often cited as the first explicit finding of a U.S. right to privacy. Warren and Brandeis wrote that privacy is the "right to be let alone" and focused on protecting individuals.[18]

On 10 December 1948, the United Nations General Assembly adopted the *Universal Declaration of Human Rights* (UDHR), originally written to guarantee the individual rights of everyone everywhere. The term Right to Privacy is not written in the document; however, many interpret this by reading Article 12, which states:

> No one shall be subjected to arbitrary interference with his privacy, family, home or correspondence, nor to attacks upon his honour and reputation. Everyone has the right to the protection of the law against such interference or attacks.[19]

Since the global surveillance disclosures of 2013, initiated by ex-NSA employee Edward Snowden, the right to privacy has been a subject of international debate. Government agencies, such as the NSA, CIA, R&AW, and GCHQ, have engaged in mass, global surveillance.

Some current debates around the right to privacy include whether privacy can co-exist with the current capabilities of intelligence agencies to access and analyze many details of an individual's life; whether or not the right to privacy is forfeited as part of the social contract to bolster defense against supposed terrorist threats; and whether threats of terrorism are a valid excuse to spy on the general population.

Private sector actors can also threaten the right to privacy—particularly technology companies, such as Amazon, Apple, Meta, Google, Microsoft, and Yahoo that use and collect personal data. These concerns have been strengthened by scandals, including the Facebook–Cambridge Analytica data scandal, which focused on the psychographic company Cambridge Analytica, which used personal data from Facebook to influence large groups of people.[20]

In Canada, the *Finestone Charter* defined the right to privacy as including physical privacy, freedom from surveillance, freedom from monitoring and interception, and freedom from collection, use, and disclosure of personal information.

---

[17] James Griffin (1 November 2007). "The Human Right to Privacy" (PDF). San Diego Law Review. p. 3.

[18] Warren, Samuel; Brandeis, Louis D. (1890). "The Right to Privacy". *Harvard Law Review*. 4: 193. Archived from the original on 23 October 2008.

[19] Nations, United. "Universal Declaration of Human Rights". United Nations.

[20] Kozlowska, Hanna. "The Cambridge Analytica scandal affected nearly 40 million more people than we thought". *Quartz*.

6.2.1.2 Freedom of Expression

Freedom of expression, or freedom of speech, has a history that predates our modern constitutional and international human rights provisions, having likely emerged during the late sixth or early fifth century BC, from the Athenian democratic principle of free speech.[21] Freedom of expression is the concept of the inherent human right to voice one's opinion publicly without fear of government censorship or punishment. In this context, the concept of "speech" is not limited to public speaking and is generally taken to include other forms of expression.

The right to freedom of speech is preserved in the United Nations *Universal Declaration of Human Rights* and is granted formal recognition by the laws of most nations. Nonetheless, the degree to which the right is upheld in practice varies greatly from one nation to another. In many nations, particularly those with authoritarian forms of government, overt government censorship is enforced. Censorship has also been claimed to occur in other forms, and there are different approaches to issues such as hate speech, obscenity, and defamation laws.[22]

Article 19 of the *Universal Declaration of Human Rights*, which was adopted in 1948, states that:

> Everyone has the right to freedom of opinion and expression; this right includes freedom to hold opinions without interference and to seek, receive, and impart information and ideas through any media and regardless of frontiers.[23]

While it is a resolution of the United Nations General Assembly, rather than a formal treaty, the *Universal Declaration of Human Rights* is not binding on members of the UN in its entirety. However, freedom of speech is granted unambiguous protection in international law by the *International Covenant on Civil and Political Rights,* which is legally binding on roughly 150 nations worldwide.[24]

In Canada, freedom of expression is included in the *Charter of Rights and Freedoms* and is protected by the *Constitution.* Freedom of expression, in the *Canadian Charter of Rights and Freedoms,* includes "freedom of thought, belief, opinion, and expression, including freedom of the press and other media of communication."

In the United States, the *Constitutional First Amendment* guarantees the rights to free expression and free association. The American Civil Liberties Union describes this as meaning that "the government does not have the right to forbid us from saying

---

[21] Raaflaub, Kurt; Ober, Josiah; Wallace, Robert (2007). *Origins of democracy in ancient Greece.* University of California Press. p. 65.

[22] Smith, David (5 February 2006). "Timeline: a history of free speech". *The Guardian.* London. Archived from the original on 29 August 2013.

[23] General Assembly of the United Nations (1948-12-10). "Universal Declaration of Human Rights" (pdf) (in English and French). pp. 4–5.

[24] Office of the United Nations High Commissioner for Human Rights. "List of Declarations and Reservations to the International Covenant on Civil and Political Rights".

what we like and writing what we like; we can form clubs and organisations and take part in demonstrations and rallies."

In Table 6.1, we can see a handful of examples of codified provisions which aim to establish and protect the right to freedom of expression, or freedom of speech.

6.2.1.3 Freedom of Assembly and Association

Freedom of assembly ensures people can gather and meet, both publicly and privately. Assemblies can be platforms to advocate for change and for people to raise awareness about the issues that matter to them, whether it is human rights, socio-economic rights, or any other issue. Freedom of association is intended to recognize the profoundly social nature of human endeavours and to protect the individual from state-enforced isolation in the pursuit of their ends.[25] It protects the collective action of individuals in pursuit of their common goals.[26] It functions to protect individuals against more powerful entities, thus empowering vulnerable groups and helping them work to right imbalances in society.[27] It allows the achievement of individual potential through interpersonal relationships and collective action.[28]

For example, in Canada's *Charter of Rights and Freedoms*:

> 2. Everyone Has the Following Fundamental Freedoms
>
> ...
>
> (d) freedom of association.[29]

Similar provisions may be found in the following Canadian laws and international instruments binding on Canada: Section 1(e) of the *Canadian Bill of Rights*; article 22 of the *International Covenant on Civil and Political Rights*; Article 8 of the *International Covenant on Economic Social and Cultural Rights*; articles 1–11 of the *International Labour Organization Convention No. 87—Freedom of Association and Protection of the Right to Organize*; article 22 of the *American Declaration of the Rights and Duties of Man*; and Article 45(c) of the Charter *of the Organization of American States.*

Similarly, although not legally binding on Canada, the following international, regional, and comparative law instruments also include similar provisions: article 11 of the *European Convention for the Protection of Human Rights and Fundamental Freedoms*; and article 16 of the *American Convention on Human Rights.*

While freedom of association is not explicitly set out in the *Constitution of the United States of America*, it has long been held to be implicit in the *First*

[25] *Mounted Police Association of Ontario v. Canada*, 2015 SCC 1 at paragraph 54.

[26] *Lavigne v. Ontario Public Service Employees Union*, [1991] 2 S.C.R. 211 at 253.

[27] *Mounted Police Association of Ontario v. Canada*, 2015 SCC 1 at paragraph 54.

[28] *Dunmore v. Ontario (Attorney General)*, [2001] 3 S.C.R. 1016 at paragraph 17.

[29] *Canadian Charter of Rights and Freedoms*, s 7, Part 1 of the Constitution Act, 1982, being Schedule B to the Canada Act 1982 (UK), 1982, c 11, s. 2(d).

Table 6.1 Codified freedom of expression/speech provisions

Entity/nation	Freedom of speech/expression provision	Source
Council of Europe	Everyone has the right to freedom of expression. This right shall include freedom to hold opinions and to receive and impart information and ideas without interference by public authority and regardless of frontiers. This article shall not prevent states from requiring the licensing of broadcasting, television or cinema enterprises. **Limitations:** The exercise of these freedoms, since it carries with it duties and responsibilities, may be subject to such formalities, conditions, restrictions, or penalties as are prescribed by law and are necessary in a democratic society, in the interests of national security, territorial integrity, or public safety, for the prevention of disorder or crime, for the protection of health or morals, for the protection of the reputation or the rights of others, for preventing the disclosure of information received in confidence, or for maintaining the authority and impartiality of the judiciary.	*European convention on human rights*[a]
European Union (EU)	Article 11: 1. Everyone has the right to freedom of expression. This right shall include freedom to hold opinions and to receive and impart information and ideas without interference by public authority and regardless of frontiers. 2. The freedom and pluralism of the media shall be respected.	*Charter of fundamental rights of the European Union*[b]
UN general assembly	"Everyone has the right to freedom of opinion and expression; this right includes freedom to hold opinions without interference and to seek, receive, and impart information and ideas through any media and regardless of frontiers."	*Universal declaration of human rights*[c]
Bangladesh	Article 39: (1) freedom of thought and conscience is guaranteed. (2) subject to any reasonable restrictions imposed by law in the interests of the security of the state, friendly relations with foreign states, public order, decency or morality, or in relation to	*Constitution of the People's republic of Bangladesh*

(continued)

Table 6.1 (continued)

Entity/nation	Freedom of speech/expression provision	Source
	contempt of court, defamation or incitement to an offence (a) The right of every citizen to freedom of speech and expression; and, (b) Freedom of the press, are guaranteed.	
Canada	Sect. 2: Everyone has the following fundamental freedoms: ... (b) Freedom of thought, belief, opinion, and expression, including freedom of the press and other media of communication **Limitations:** (1) the Canadian charter of rights and freedoms guarantees the rights and freedoms set out in it subject only to such reasonable limits prescribed by law as can be demonstrably justified in a free and democratic society.	*Canadian charter of rights and freedoms*[d]
China	Article 35: Citizens of the People's Republic of China enjoy freedom of speech, of the press, of assembly, of association, of procession, and of demonstration. **Limitations:** No unit or individual may use the Internet to create, replicate, retrieve, or transmit the following kinds of information: 1. Inciting to resist or violate the constitution or laws or the implementation of administrative regulations 2. Inciting to overthrow the government or the socialist system 3. Inciting division of the country, harming national unification 4. Inciting hatred or discrimination among nationalists or harming the unity of the nationalities 5. Making falsehoods or distorting the truth, spreading rumours, destroying the order of society 6. Promoting feudal superstitions, sexually suggestive material, gambling, violence, murder 7. Engaging in terrorism or inciting others to criminal activity; openly	"Chap. II". *Constitution of the People's republic of China*[f]

(continued)

Table 6.1 (continued)

Entity/nation	Freedom of speech/expression provision	Source
	insulting other people or distorting the truth to slander people 8. Injuring the reputation of state organs 9. Other activities against the constitution, laws or administrative regulations[e]	
Cuba	Article 53: Citizens have freedom of speech and of the press in keeping with the objectives of socialist society. Material conditions for the exercise of that right are provided by the fact that the press, radio, television, movies, and other organs of the mass media are state or social property and can never be private property. This assures their use at the exclusive service of the working people and in the interest of society. The law regulates the exercise of these freedoms. Article 55: The rights of assembly, demonstration, and association are exercised by workers, both manual and intellectual; peasants; women; students; and other sectors of the working people, [rights] to which they have the necessary ability (los medios necesarios) to exercise. The social and mass organisations have all the facilities they need to carry out those activities in which the members have full freedom of speech and opinion based on the unlimited right of initiative and criticism.	*Constitution of Cuba*[g]
Czech Republic	Article 17: 1. The freedom of expression and the right to information are guaranteed. 2. Everyone has the right to express their opinion in speech, in writing, in the press, in pictures, or in any other form, as well as freely to seek, receive, and disseminate ideas and information irrespective of the frontiers of the state. 3. Censorship is not permitted. 4. The freedom of expression and the right to seek and disseminate information *may be limited by law in the case of measures necessary in a democratic society for protecting the rights and freedoms of others, the security of the state, public security, public health, and*	*Charter of fundamental rights and basic freedoms the Czech Republic*[t] *Criminal code of the Czech Republic*[u]

(continued)

Table 6.1 (continued)

Entity/nation	Freedom of speech/expression provision	Source
	morals. 5. State bodies and territorial self-governing bodies are obliged, in an appropriate manner, to provide information on their activities. Conditions therefore and the implementation thereof shall be provided for by law. **Limitations:** • Unauthorised handling of personal information[h] – Protects the right to privacy • Defamation[i] • Dissemination of pornography depicting disrespect to a human, abuse of an animal, or dissemination of any pornography to children[j] • Seducing to use or propagation of use of addictive substances other than alcohol[k] – Protects public health • Denigration of a nation, race, ethnicity, or other group of people[l] – i.e. hate speech • Inciting of hatred towards a group of people or inciting limitation of their civil rights[m] • Spreading of scaremongering information[n] – e.g. fake bomb alerts • Public incitement to the perpetration of a crime[o] • Public approval of a felony crime[p] • Public display of sympathy towards a movement oriented at curbing rights of the people[q] – e.g. propagation of hate groups • Public denial, questioning, endorsement, or vindication of genocide[r] – e.g. Auschwitz lie • Incitement to an offensive war[s]	
Ecuador	Article 66, Subsection 6: The right to voice one's opinion and express one's thinking freely and in all of its forms and manifestations	*Constitution of Ecuador, 2008*[v]
Hong Kong	Article 27: Hong Kong residents shall have freedom of speech, of the press and of publication; freedom of association, of assembly, of procession, and of	"Chap. III: Fundamental rights and duties of the residents". *The basic law of the Hong Kong special administrative Region of the People's republic of China*[w]

(continued)

Table 6.1 (continued)

Entity/nation	Freedom of speech/expression provision	Source
	demonstration; and the right and freedom to form and join trade unions, and to strike. Article 30: The freedom and privacy of communication of Hong Kong residents shall be protected by law. No department or individual may, on any grounds, infringe upon the freedom and privacy of communication of residents except that the relevant authorities may inspect communication in accordance with legal procedures to meet the needs of public security or of investigation into criminal offences.	
India	Article 19: All citizens shall have the right: 1. To freedom of speech and expression 2. To assemble peaceably and without arms 3. To form associations or unions 4. To move freely throughout the territory of India 5. To reside and settle in any part of the territory of India 6. To practise any profession, or to carry on any occupation, trade, or business **Limitations:** These rights are limited so as not to affect: – The integrity of India – The security of the state – Friendly relations with foreign states – Public order – Decency or morality – Contempt of court – Defamation or incitement to an offence	"19. Protection of certain rights regarding freedom of speech, etc". *The constitution of India*.[x]
Japan	Article 21: Freedom of assembly and association as well as speech, press, and all other forms of expression are guaranteed No censorship shall be maintained, nor shall the secrecy of any means of communication be violated	"Chap. III". *Constitution of Japan*[y]
New Zealand	Sect. 14: Everyone has the right to freedom of expression, including the freedom to	*New Zealand Bill Of Rights Act 1990*[z]

(continued)

Table 6.1 (continued)

Entity/nation	Freedom of speech/expression provision	Source
	seek, receive, and impart information and opinions of any kind in any form	
Peru	Article 2, Section 4: To freedom of information, opinion, expression, and dissemination of thought, whether oral, written, or in images, through any medium of social communication, and without previous authorization, censorship, or impediment, under penalty of law Crimes committed by means of books, the press, and any other social media are defined by the criminal code and tried in a court of law Any action that suspends or closes down any organ of expression or prevents its free circulation constitutes a crime. The rights of information and opinion include those of founding means of communication	*Political constitution of Peru*[aa]
Philippines	Article III, Section 4: No law shall be passed abridging the freedom of speech or of expression	*Constitution of the Philippines*[ab]
South Africa	Sect. 16: Everyone has the right to freedom of expression, which includes: (a) Freedom of the press and other media (b) Freedom to receive or impart information or ideas (c) Freedom of artistic creativity (d) Academic freedom and freedom of scientific research **Limitations:** This right does not extend to: (a) Propaganda for war (b) Incitement of imminent violence (c) Advocacy of hatred that is based on race, ethnicity, gender, or religion and that constitutes incitement to cause harm	*South African bill of rights*[ac]
United Kingdom	Article 10: Everyone has the right to freedom of expression. This right shall include freedom to hold opinions and to receive and impart information and ideas without interference by public authority and regardless of frontiers. This article shall	*The Human Rights Act 1998*

(continued)

Table 6.1 (continued)

Entity/nation	Freedom of speech/expression provision	Source
	not prevent states from requiring the licensing of broadcasting, television or cinema enterprises **Limitations:** The exercise of these freedoms, since it carries with it duties and responsibilities, may be subject to such formalities, conditions, restrictions, or penalties as are prescribed by law and are necessary in a democratic society, in the interests of national security, territorial disorder, or crime, for the protection of health or morals, for the protection of the reputation or rights of others, for preventing the disclosure of information received in confidence, or for maintaining the authority and impartiality of the judiciary	
United States	Congress shall make no law respecting an establishment of religion, or prohibiting the free exercise thereof; or abridging the freedom of speech, or of the press; or the right of the people peaceably to assemble, and to petition the government for a redress of grievances	*First amendment to the United States constitution*[ad]

[a] *European Convention on Human Rights* (1950)
[b] *Charter of Fundamental Rights of the European Union* (2000) (2007/C 303/01)
[c] General Assembly of the United Nations. *Universal Declaration of Human Rights* (1948). pp. 4–5
[d] *Canadian Charter of Rights and Freedoms*, s 7, Part 1 of the Constitution Act, 1982, being Schedule B to the Canada Act 1982 (UK), 1982, c 11
[e] Article 5 of the *Computer Information Network and Internet Security, Protection and Management Regulations,* 1997
[f] "Chap. II. The Fundamental Rights and duties of Citizens". *Constitution of the People's Republic of China*. (1982), c 2, article 35
[g] "The Constitution of the Republic of Cuba, 1976 (as Amended to 2002)" (PDF). *constitutionnet.org*. 2002
[h] "Criminal Code of the Czech Republic, §180". *Collection of the Laws of the Czech Republic* (in Czech). Prague. 40 (2009)
[i] "Criminal Code of the Czech Republic, §184". *Collection of the Laws of the Czech Republic* (in Czech). Prague. 40 (2009)
[j] "Criminal Code of the Czech Republic, §191". *Collection of the Laws of the Czech Republic* (in Czech). Prague. 40 (2009)
[k] "Criminal Code of the Czech Republic, §287". *Collection of the Laws of the Czech Republic* (in Czech). Prague. 40 (2009)
[l] "Criminal Code of the Czech Republic, §355". *Collection of the Laws of the Czech Republic* (in Czech). Prague. 40 (2009)
[m] "Criminal Code of the Czech Republic, §356". *Collection of the Laws of the Czech Republic* (in Czech). Prague. 40 (2009)

(continued)

Table 6.1 (continued)

[n] "Criminal Code of the Czech Republic, §357". *Collection of the Laws of the Czech Republic* (in Czech). Prague. 40 (2009)
[o] "Criminal Code of the Czech Republic, §364". *Collection of the Laws of the Czech Republic* (in Czech). Prague. 40 (2009)
[p] "Criminal Code of the Czech Republic, §365". *Collection of the Laws of the Czech Republic* (in Czech). Prague. 40 (2009)
[q] "Criminal Code of the Czech Republic, §404". *Collection of the Laws of the Czech Republic* (in Czech). Prague. 40 (2009)
[r] "Criminal Code of the Czech Republic, §405". *Collection of the Laws of the Czech Republic* (in Czech). Prague. 40 (2009)
[s] "Criminal Code of the Czech Republic, §407". *Collection of the Laws of the Czech Republic* (in Czech). Prague. 40 (2009)
[t] Czech National Council (16 December 1992). "Charter of Fundamental Rights and Basic Freedoms the Czech Republic". *Collection of the Laws of the Czech Republic*. Prague. 2 (1993)
[u] "Criminal Code of the Czech Republic". *Collection of the Laws of the Czech Republic* (in Czech). Prague. 40 (2009)
[v] "Ecuador: 2008 Constitution in English". *pdba.georgetown.edu*
[w] "Chap. III: Fundamental Rights and Duties of the Residents". *The Basic Law of the Hong Kong Special Administrative Region of the People's Republic of China*. July 13, 2012
[x] "19. Protection of certain rights regarding freedom of speech, etc". *The Constitution of India* (1949) p. 8
[y] "Chap. III". *Constitution of Japan* (1946)
[z] *New Zealand Bill of Rights Act 1990*, s 14
[aa] "Political Constitution of Peru". *Congresso de Peru*. October 31, 1993. Article 2, Section 4
[ab] Article III, Section 4, *Constitution of the Philippines* (1987)
[ac] "Bill of Rights" (1996). In the *Constitution of the Republic of South Africa*, c 2, s 16
[ad] US Const. amend I

Amendment's protection of freedom of speech, assembly, and petition. With respect to collective bargaining, see *International Labour Organization Convention No. 98* concerning the application of the principles of the right to organise and to bargain collectively.

Freedom of association protects three classes of activities: (1) the "constitutive" right to join with others and form associations; (2) the "derivative" right to join with others in the pursuit of other constitutional rights; and (3) the "purposive" right to join with others to meet on more equal terms the power and strength of other groups or entities. Under the constitutive right, the state is prohibited from interfering with individuals meeting or forming associations but is permitted to interfere with the *activities* pursued by an association. The derivative right protects associations' activities that specifically relate to other constitutional freedoms but does not protect other activities of the association. The purposive right protects associations' activities, including collective bargaining and striking, that enable individuals who are vulnerable and ineffective to meet on more equal terms the power and strength of those with whom their interests interact or conflict.[30]

[30] *Mounted Police Association of Ontario v. Canada*, 2015 SCC 1 at paragraphs 52–54, 66.

Sect. 2(d) also encompasses what has been called a “negative aspect”, a “freedom not to associate” or a “freedom from compelled (or ‘forced’) association”. However, Section 2(d) is not a constitutional right to isolation. It does not protect against all forms of involuntary association and was not intended to protect against association with others that is a necessary and inevitable part of membership in a modern democratic community.[31] Compelled association in the form of legal obligations arising from these unavoidable types of associations does not in and of itself offend Sect. 2(d).[32,33]

Forced association threatens an identified liberty interest when there is imposition of a form of ideological conformity on the claimant; government establishment of, or support for, particular political causes; impairment of an individual’s freedom to join or associate with causes of his or her choosing; and personal identification of an individual with causes which he or she does not support.[34]

The United Nations recognizes the importance of the rights to peaceful assembly and of association to the full enjoyment of civil and political rights and economic, social, and cultural rights. In October 2010, the Human Rights Council adopted resolution 15/21 establishing the mandate of the Special Rapporteur on the rights to freedom of peaceful assembly and of association for an initial period of three years. This mandate was created to:

- Gather and share information about global, regional, and local trends and issues relating to peaceful assembly and association.
- Make recommendations on how to ensure the promotion and protection of these rights.
- Report on violations, as well as discrimination, threats or use of violence, harassment, persecution, intimidation, or reprisals directed at persons exercising these rights.

The Council extended the mandate of the Special Rapporteur for the first time in September 2013 (resolution 24/5) and a second time in June 2016 (resolution 32/32). The mandate was last renewed in July 2019 (resolution 41/12) for a period of three years. The mandate-holder serves for an initial period of three years, renewable once.

6.2.1.4 Right to (Digital) Access, Education, and Development

Several countries have adopted laws that require the state to work to ensure that Internet access is broadly available and/or prevent the state from unreasonably restricting an individual’s access to information and the Internet:

---

[31] *Bernard v. Canada*, [2014] 1 S.C.R. 227 at paragraph 38.

[32] *Lavigne v. Ontario Public Service Employees Union*, [1991] 2 S.C.R. 211 at 320–321.

[33] *R. v. Advance Cutting & Coring Ltd.*, [2001] 3 S.C.R. 209, at paragraphs 19, 194, 223, 232.

[34] *R. v. Advance Cutting & Coring Ltd.*, [2001] 3 S.C.R. 209.

- **Costa Rica**

  A 30 July 2010 ruling by the Supreme Court of Costa Rica stated, "Without fear of equivocation, it can be said that these technologies [information technology and communication] have impacted the way humans communicate, facilitating the connection between people and institutions worldwide and eliminating barriers of space and time. At this time, access to these technologies has become a basic tool to facilitate the exercise of fundamental rights and democratic participation (e-democracy) and citizen control, education, freedom of thought and expression, access to information and public services online, the right to communicate with government electronically and administrative transparency, among others. This includes the fundamental right of access to these technologies, in particular, the right of access to the Internet or World Wide Web."[35]
- **Estonia**

  In 2000, the parliament launched a massive program to expand access to the countryside. The Internet, the government argues, is essential for life in the twenty-first century.[36]
- **European Union**

  As of April 2022, the European Parliament is considering a *Declaration on Digital Rights*.[37]
- **Finland**

  By July 2010, every person in Finland was to have access to a one-megabit-per-second broadband connection, according to the Ministry of Transport and Communications. And by 2015, access to a 100 Mbit/s connection.[38]
- **France**

  In June 2009, the Constitutional Council, France's highest court, declared access to the Internet to be a basic human right in a strongly worded decision that struck down portions of the HADOPI law, a law that would have tracked abusers without judicial review and automatically cut off network access to those who continued to download illicit material after two warnings.[39]
- **Greece**

  Article 5A of the *Constitution of Greece* states that all persons have the right to participate in the Information Society and that the state has an obligation to facilitate the production, exchange, diffusion, and access of electronically transmitted information.[40]
- **Spain**

[35] "Judgement 12,790 of the Supreme Court" Archived 2015-12-17.

[36] "Estonia, where being wired is a human right", Colin Woodard, *Christian Science Monitor*, 1 July 2003.

[37] European Commission (26 January 2022). *European Declaration on Digital Rights and Principles for the digital decade — COM(2022) 28 final.* Brussels, Belgium: European Commission.

[38] "Finland makes 1 Mb broadband access a legal right", Don Reisinger, *CNet News*, 14 October 2009.

[39] "Top French Court Declares Internet Access 'Basic Human Right'". *London Times*. Fox News. 12 June 2009.

[40] *Constitution of Greece As revised by the parliamentary resolution of May 27th 2008 of the VIIIth Revisionary Parliament*, English language translation, Hellenic Parliament.

Starting in 2011, Telefónica, the former state monopoly that holds the country's "universal service" contract, has to guarantee to offer "reasonably" priced broadband of at least one megabyte per second throughout Spain.[41]

6.2.2 Legal Rights

Legal rights are those which exist under the rules of legal systems or by virtue of decisions of suitably authoritative bodies within them. Their use is pervasive in modern legal systems. Legal rights are different from a moral or natural right in the sense that legal rights are recognized and protected by law, whereas moral rights may not be explicitly recognized and protected by law. A court of law can enforce legal rights against individuals, corporations, and government, as a legal right is an interest accepted and protected by law. Any debasement of any legal right is equally punishable by the law.

Table 6.2 describes the different types of legal rights, with examples provided for each.

We can look to the *Canadian Charter of Rights and Freedoms*, as it applies within Canada, for an example of how moral rights and legal rights interact and are protected under the *Charter*. While many countries have different protected rights and freedoms, Canada is one example of how a nation has compiled these rights and freedoms into a "living" protective document, that being the Canadian *Constitution*.[42]

6.2.2.1 Example: Canadian Legal Rights

In Canada, Section 7 of the *Charter of Rights and Freedoms* protects substantive legal rights from restriction. This section ensures that our rights to "life, liberty, and security of the person" are only infringed by laws that conform to the "principles of fundamental justice."[43] There are three established principles of fundamental justice. A law that restricts section. 7 rights must not be arbitrary: it must relate to a specific problem. The law must not be overbroad: it must only restrict section. 7 rights to the extent necessary. Finally, a law must not be disproportionate: the restriction on rights must not outweigh the problem the law addresses.

---

[41] Sarah Morris (17 November 2009). "Spain govt to guarantee legal right to broadband". Reuters.

[42] *Canadian Charter of Rights and Freedoms*, Part I of the *Constitution Act, 1982*, being Schedule B to the *Canada Act 1982* (UK), 1982, c 1.

[43] *Canadian Charter of Rights and Freedoms*, Part I of the *Constitution Act, 1982*, being Schedule B to the *Canada Act 1982* (UK), 1982, c 1.

Table 6.2 Types of legal rights

Type of rights	Description	Example(S)
Perfect rights	Rights which are recognized by law and enforceable by law	Constitutionally-protected fundamental rights
Imperfect rights	Rights which are recognized by law but which are not enforceable by law	Time-bound claims or debts
Positive rights	Requires the performance of some act, corresponding with a positive duty to fulfil the right	Right to education, housing, medical care, etc.
Negative rights	Prevents the performance of some act, corresponding with a negative duty to fulfil the right	Right to life, liberty, security, etc.
Rights in rem	Rights which are based on duties imposed upon others and available against the whole world, in general	Tort or criminal laws
Rights in persona	Rights which are based on a duty imposed upon a particular person and available against a particular person only	Contractual obligations, breach of contract, etc.
Proprietary rights	Rights which are available with respect to a property, as it is related to the owner and the assets of the owner (inherent monetary value)	Right to property ownership, right to patent, ownership rights, etc.
Personal rights	Rights which are related to the life, reputation, or social standing of a particular person (no inherent monetary value)	Right to life
Public rights	Rights which are vested in a person by a state, government, or constitution	Right to vote
Private rights	Rights which are connected to private individuals or entities	Contractual rights
Inheritable rights	Rights which can be passed from one generation to another, surviving even after the death of the original right holder	A legal heir to a property
Uninheritable rights	Rights which cannot be passed from one generation to another and which die with the death of the original right holder	All personal rights
Rights in Repropria	Rights with respect to one's absolute ownership over one's own property	Right to use, dispose of, destroy, modify, or exclude from others
Rights in Realiena	Rights with respect to the property of another person, not absolute	Right of way, or easement, over the property of another

Legal rights ensure the state treats citizens fairly. Section 7 specifically ensures that laws do not unfairly restrict the life, liberty, or security of people living in Canada. Sections 8 through 14 protect people in their interactions with the justice system and law enforcement.[44]

[44] *Canadian Charter of Rights and Freedoms*, Part I of the *Constitution Act, 1982*, being Schedule B to the *Canada Act 1982* (UK), 1982, c 1.

Sections 8 through 14 of the *Charter of Rights and Freedoms* protect procedural legal rights.[45] The legal rights section of the *Charter* guarantees the rights held by anyone investigated, detained, or criminally charged, or who is faced with legal restrictions of any kind.

These rights explain what state agents such as police officers may do when investigating, or prosecuting, a legal infraction. As "the treatment meted out by agents of the state to even the least deserving individual will often indicate the treatment that all citizens of the state may ultimately expect," courts will diligently protect procedural rights. These include Section 8, which protects against unreasonable search and seizure, Section 9, which protects against arbitrary detention, and Section 12, which protects against cruel and unusual punishment. Sections 10, 11, and 13 protect rights upon and after arrest, such as the right to legal counsel and the right to trial within a reasonable time.

We can summarise all of the rights of Canadians—moral, legal, and equality-based—as per the *Canadian Charter of Rights and Freedoms,* as follows:

Fundamental Freedoms

Everyone in Canada is free to practise any religion or no religion at all. We are also free to express religious beliefs through prayer or by wearing religious clothing, for example. However, the *Charter* also ensures that others have the right to express their religious beliefs in public.

We are free to think our own thoughts, speak our minds, listen to the views of others, and express our opinions in creative ways. We are also free to meet with anyone we wish and participate in peaceful demonstrations. This includes the right to protest a government action or institution.

However, these freedoms are not unlimited. There may be limits on how you express your religious beliefs if your way of doing so would infringe on the rights of others or undermine complex public programs and policies. For example, you may have religious reasons to object to having your photo taken for your driver's licence, but this requirement may be linked to a need to stop others from unlawfully using your identity. In addition, the *Charter* does not protect expression such as hate speech that involves threats of violence or that takes the form of violence.

The media also have certain fundamental freedoms and are free to print and broadcast news and other information. The government can only limit what the media prints for justifiable reasons set out in law. For example, a magazine cannot print slander, which is an untrue statement about a person that may hurt his or her reputation.

[45] *Canadian Charter of Rights and Freedoms*, s 7, Part 1 of the Constitution Act, 1982, being Schedule B to the Canada Act 1982 (UK), 1982, c 11.

Democratic Rights

Every Canadian citizen has the right to vote in elections and to run for public office themselves. There are certain exceptions. For example, people must be 18 years old or older to vote.

Our elected governments cannot hold power for an unlimited amount of time. The *Charter* requires governments to call an election at least once every five years. An election could be delayed, however, during a national emergency, like a war. In this case, two-thirds of the members of the House of Commons or, in the case of provinces or territories, the legislative assembly must agree to delay the election. The *Charter* makes it clear that elected representatives of legislative assemblies must sit at least once a year. This holds Parliament and all other legislatures responsible for their actions.

Mobility Rights

Canadian citizens have the right to enter, remain in, and leave Canada. Canadian citizens and permanent residents have the right to live or seek work anywhere in Canada. Governments in Canada can't discriminate based on what province someone used to live in or currently lives in.

However, laws can set certain rules for when people are able to get social, health, and welfare benefits. For example, you may have to live in a particular province for a certain length of time before getting health benefits from that province. Also, provinces with an employment rate below the national average may create programs that are only available to their own socially and economically disadvantaged residents. These programs encourage residents to stay in the province and contribute to the workforce.

Legal Rights

1. **Right to be secure against unreasonable search and seizure**

The *Charter* protects everyone's reasonable expectation of privacy. This means that no one can search you, take away your personal belongings, or access your personal information without clear legal reasons.

Authorities acting on behalf of the government, such as the police, must carry out their duties in a fair and reasonable way. For example, they cannot enter private property or take things without good reason. Police are required to get a warrant from a judge before searching someone's home.

2. **Interactions with the justice system**

The *Charter* sets out rights that deal with the interaction between the justice system and individuals. These rights ensure that individuals are treated fairly at every stage of the justice process. This is especially true if an individual is charged with a criminal offence.

3. **Protection against unreasonable laws**

The *Charter* protects everyone against unreasonable laws that could lead to imprisonment or harm their physical safety. The law may still comply with the *Charter* if it is consistent with a basic set of values. For example, there must be a rational link between the law's purpose and its effect on people's liberty. Also, laws should not have a severe impact on people's rights to life, liberty, or security of the person.

4. **Protection against arrest without good reason**

The *Charter* also says law enforcement agencies cannot take actions against individuals that are random or not backed by good reasons. A police officer, for example, must have reasonable grounds to believe you have committed a crime and must tell you why you are being arrested and detained. You also have the right to consult a lawyer without delay and to be informed of this right. Finally, you have the right to have a court decide whether this detention is lawful. If you believe your detention is not legal, the *Charter* protects your right to challenge it.

5. **Rights after arrest**

If you are charged with an offence under federal or provincial law, you have the right to:

- Be told quickly of the offence you are charged with
- Be tried within a reasonable amount of time
- Choose not to testify at your own trial
- Be presumed innocent until proven guilty beyond a reasonable doubt in a fair and public hearing by an independent and unbiased tribunal
- Only be denied reasonable bail with cause
- Be tried by a jury for serious charges
- Be convicted only for an act or omission that was a crime at the time it was committed
- Only be tried or punished once for an offence
- Receive the lesser punishment if the punishment for the crime changes between the time it was committed and the time of sentencing

6. **Protection against cruel and unusual punishment**

In addition, the *Charter* protects everyone from cruel and unusual punishment. This includes torture and the excessive or abusive use of force by law enforcement officials. Also, sentences of imprisonment must match the seriousness of the crime committed. For example, an extremely long prison sentence is not appropriate for a very minor crime.

7. **Rights in Court**

The *Charter* offers certain protections if you are accused of a crime and must go to court. This includes your right to a quick and reasonably speedy trial. This trial must be fair and conducted by an unbiased court that assumes your innocence until you're proven guilty. You are also entitled to an interpreter during court proceedings if you do not understand the language or if you are hearing impaired.

Anyone who is a witness in a trial has the right to not have incriminating evidence used against them in later proceedings. For example, if you admit to a crime while acting as a witness in court at someone else's trial, the police cannot use it to prove your guilt in court later. Perjury, which is lying during legal proceedings, is the one exception to this rule. The *Youth Criminal Justice Act* protects people under the age of 18.[46]

Equality Rights

Equality rights are at the core of the *Charter.* They are intended to ensure that everyone is treated with the same respect, dignity, and consideration (i.e. without discrimination), regardless of personal characteristics such as race, national or ethnic origin, colour, religion, sex, age, mental or physical disability, sexual orientation, residency, marital status, or citizenship.

As a result, everyone should be treated the same under the law. Everyone is also entitled to the same benefits provided by laws or government policies. However, the *Charter* does not require the government to always treat people in the same way. Sometimes protecting equality means that we must adapt rules or standards to take account of people's differences. An example of this would be allowing people to observe different religious holidays without losing their jobs.

Governments can also promote equality by passing laws or creating programs that aim to improve the conditions of people who have been disadvantaged because of the personal characteristics listed above. For example, governments can create affirmative action programs targeted at increasing employment for persons with disabilities.

Official Language Rights

The *Charter* establishes that English and French are the official languages of the country and have equality of status and equal rights and privileges as to their use in all institutions of the Parliament and government of Canada. It also establishes the principle of advancement or progression of the equality of status and use of the official languages by Parliament or by the provincial legislatures. Parliament acted

---

[46] https://www.justice.gc.ca/eng/cj-jp/yj-jj/tools-outils/back-hist.html

on this principle of advancement in 1988 by passing the *Official Languages Act* and by providing, in the *Criminal Code*, for the right to a trial in the official language of the accused. Many provincial and territorial governments have also passed language legislation.

The *Charter* establishes that everyone has the right to use English or French in any debates and other proceedings of Parliament. The statutes, records, and journals of Parliament must be printed and published in both languages, and both language versions are equally authoritative. Everyone has the right to use English or French in, or in any pleading in, or process issuing from, any court established by Parliament.

Any member of the public also has the right to communicate with and receive services in English or French from any head office of an institution of Parliament or the government of Canada. They have this same right from any office of an institution where there is a significant demand for communications with and services from that office in such language, or where due to the nature of the office, it is reasonable that communications with and services from that office be available in both English and French.

Similar rights apply in New Brunswick, the only officially bilingual province in Canada. In fact, members of the public in New Brunswick have the right to communicate with and obtain services in either English or French from any office of an institution of the legislature or government of New Brunswick. The English and French linguistic communities in New Brunswick also have equality of status and equal rights and privileges. This includes the right to distinct educational institutions and cultural institutions that preserve and promote those communities.

Minority Language Educational Rights

Every province and territory has official language minority communities (French-speaking communities outside Quebec and English-speaking minorities in Quebec). Sect. 23 of the *Charter* guarantees minority language educational rights to French-speaking communities outside Quebec and to English-speaking minorities in Quebec. It applies to all provinces and territories.

Canadian citizens living outside of Quebec have the right to send their children to French-speaking schools if:

- Their mother tongue is French.
- They attended French-speaking primary and secondary schools in Canada.
- They have child who has attended or is attending French-speaking primary or secondary schools in Canada.

Canadian citizens living in Quebec have the right to send their children to English-speaking schools if:

- They themselves attended English-speaking primary and secondary schools in Canada.

- They have a child who has attended or is attending English-speaking primary or secondary schools in Canada.

This right to minority-language instruction applies where there is a large enough number of people to justify it.

6.2.3 *Indigenous Rights*

Indigenous rights are inherent, collective rights that flow from the original occupation of the land, and from social orders created before the arrival of colonising groups. For many, the concept of Indigenous rights can be summed up as the right to independence through self-determination regarding governance, land, resources, and culture.

6.2.3.1 Example: Indigenous Rights in Canada

In Canada, it has been argued that it is difficult to establish a simple definition of Indigenous rights with respect to Canada because of the diversity among Indigenous peoples. For example, First Nations that have signed treaties with the federal government may enjoy certain privileges (such as annual cash payments) that non-treaty nations do not. Similarly, Indigenous nations that have won court cases regarding land claims may exercise more control over their lands and populations than others. In general, however, all Indigenous peoples have rights that may include access to ancestral lands and resources and the right to self-government.

In addition to treaties, which are supposed to enshrine certain rights to land, resources, and more, federal law also protects Indigenous rights, namely the *Constitution Act, 1982*. Since 2008, the rights of First Nations people living on reserves have also been covered by the *Canadian Human Rights Act*. Supreme Court cases have clarified definitions of Indigenous rights, and particularly Indigenous rights (or title) to traditional territories. For example, the *Delgamuukw* case in 1997 showed that Aboriginal title constituted an ancestral right protected by the *Constitution*.

The *Indian Act*—another federal law—does not enshrine rights (quite the contrary, it has been historically oppressive), but it has impacted Indigenous rights. The *Indian Act* creates legal categories of status and non-status Indians that have caused division among Indigenous peoples. For example, Status Indians have certain rights that non-status Indians do not, such as the right to not pay federal or provincial taxes on certain goods and services while living or working on reserves. However, many Indigenous peoples (both status and non-status) refuse to be defined by this federal law.

Indigenous rights are upheld and challenged at the provincial and local levels as well. Many First Nations have signed land claim agreements with the federal and provincial governments. When rights to territory are challenged, relations between these groups become less amicable. The Oka crisis and Ipperwash crisis are but two instances where provincial and local authorities ignored Indigenous claims to ancestral lands. Since the arrival of Europeans, Indigenous peoples have had to protect their rights, lands, peoples, and ways of life.

On the national stage, the *Royal Proclamation of 1763* has historically been viewed as the constitutional basis for Indigenous treaties and a source of legal rights. Affirmed by Sect. 35 of the *Constitution Act*, 1982, the legal principles of the *Royal Proclamation* are still applied in modern-day treaties.

The inclusion of Sect. 35 in the *Constitution* signalled a new era of judicial and political opinion on the question of Indigenous rights. This section protects a spectrum of different Indigenous and treaty rights, including legal recognition of customary practices such as marriage and adoption, the site-specific exercise of food harvesting and other rights that do not involve claims to the land itself, and assertions of ownership of traditional lands.

The courts, and more specifically, the Supreme Court of Canada, have clarified and guaranteed rights to land and resource activities as well as other issues. Since governments could not come to a consensus during constitutional negotiations about Indigenous rights, the issue was consequently left to the courts. The rulings become part of Canadian law and may alter the way that the government understands Indigenous rights. For example, the *Calder* case helped set the stage for many First Nations in British Columbia to launch their own land claims and cases relating to Aboriginal title.

The content and priority of issues surrounding Indigenous rights and title continue to evolve judicially and through the negotiation and implementation of self-government agreements between Indigenous peoples and the Government of Canada. Over the long term, it is likely that these issues will need to be addressed through negotiated political resolution.

On the international stage, Inuit peoples have had their rights to Arctic lands and waters affirmed by such declarations as the *United Nations Declaration on the Rights of Indigenous Peoples*. As well, Indigenous groups have participated in United Nations working groups concerned with Indigenous populations and minority rights. Although most nations adopted the UN *Declaration on the Rights of Indigenous Peoples* in 2007—an agreement that recognizes Indigenous rights to self-government, land, equality, and language, as well as basic human rights—Canada only signed on in May 2016 after a change in the federal government. Canada initially refused to sign because of issues concerning land disputes and the declaration's clauses about the duty to consult that could impact resource development. It has yet to be seen how Canada will implement this agreement.

6.2.3.2 United Nations Declaration on the Rights of Indigenous Peoples

The *United Nations Declaration on the Rights of Indigenous Peoples* (UNDRIP) was adopted by the General Assembly on Thursday, 13 September 2007, by a majority of 144 states in favour, 4 votes against, and 11 abstentions.[47]

The four votes against were made by Australia, Canada, New Zealand, and the United States. The nations which abstained were Azerbaijan, Bangladesh, Bhutan, Burundi, Colombia, Georgia, Kenya, Nigeria, Russian Federation, Samoa and Ukraine.

Eventually, the four countries that voted against the *Declaration*—Australia, Canada, New Zealand, and the United States—reversed their position and now support the UN *Declaration*. Today the *Declaration* is the most comprehensive international instrument on the rights of indigenous peoples. It establishes a universal framework of minimum standards for the survival, dignity, and well-being of the indigenous peoples of the world and it elaborates on existing human rights standards and fundamental freedoms as they apply to the specific situation of indigenous peoples.

The *Declaration* is a comprehensive statement addressing the human rights of indigenous peoples. It was drafted and formally debated for over twenty years prior to being adopted by the General Assembly on 13 September 2007. The document emphasises the rights of indigenous peoples to live in dignity, to maintain and strengthen their own institutions, cultures, and traditions, and to pursue their self-determined development, in keeping with their own needs and aspirations. Other UN bodies address indigenous rights through conventions such as the *International Labour Organization's Convention No.169* and the *Convention on Biological Diversity* (Article 8j).

The *Declaration* addresses both individual and collective rights, cultural rights and identity, rights to education, health, employment, language, and others. The text says indigenous peoples have the right to fully enjoy as a collective or as individuals, all human rights and fundamental freedoms as recognized in the *Charter of the United Nations*, the *Universal Declaration of Human Rights*, and the rest of international human rights law. Indigenous peoples and individuals are free and equal to all other peoples and individuals and have the right to be free from any kind of discrimination, in the exercise of their rights, in particular that based on their indigenous origin or identity. Indigenous peoples have the right to self-determination. By that right they can freely determine their political status and pursue their economic, social, and cultural development. They have the right to maintain and strengthen their distinct political, legal, economic, social, and cultural institutions while retaining their right to participate fully, if they choose to, in the political, economic, social, and cultural life of the state.

Notably, seventeen of the forty-six articles of the *Declaration* deal with indigenous culture and how to protect and promote it, by respecting the direct input of

[47] *United Nations Declaration on the Rights of Indigenous Peoples* (A/RES/61/295).

indigenous peoples in decision-making and allowing for resources, such as those for education in indigenous languages and other areas. Fifteen of the articles of the *Declaration* are about indigenous peoples' participation in all decisions that will affect their lives, including meaningful participation in a democratic polity. The *Declaration* confirms the right of indigenous peoples to self-determination and recognizes subsistence rights and rights to lands, territories, and resources. The *Declaration* recognizes that indigenous peoples deprived of their means of subsistence and development are entitled to just and fair redress. Essentially, the *Declaration* outlaws discrimination against indigenous peoples, promotes their full and effective participation in all matters that concern them, as well as their right to remain distinct and to pursue their own visions of economic and social development.

While UN *Declarations* are generally not legally binding, they represent the dynamic development of international legal norms and reflect the commitment of states to move in certain directions, abiding by certain principles. The *Declaration*, however, is widely viewed as not creating new rights. Rather, it provides a detailing or interpretation of the human rights enshrined in other international human rights instruments of universal resonance—as these apply to indigenous peoples and indigenous individuals. It is in that sense that the *Declaration* has a binding effect for the promotion, respect, and fulfilment of the rights of indigenous peoples worldwide. The *Declaration* is a significant tool towards eliminating human rights violations against the over 370 million indigenous people worldwide and assisting them and States in combating discrimination and marginalization.

While it is not viewed as creating new rights, many of the rights provided in the *Declaration* do require new approaches to global issues, such as development, decentralization, and multicultural democracy. In order to achieve full respect for diversity, countries will need to adopt participatory approaches to indigenous issues, which will require effective consultations and the building of partnerships with indigenous peoples.

6.3 Indigenous Data Sovereignty

Indigenous data sovereignty has emerged as an important topic, raising fundamental questions about assumptions of ownership, representation, and control in open data communities. Indigenous data sovereignty refers to the right of Indigenous peoples to control data from and about their communities and lands, articulating both individual and collective rights to data access and to privacy. It is the ability for Indigenous peoples, communities, and nations to participate, steward, and control data that is created by or about themselves.

The term sovereignty refers to the fact that Indigenous nations are sovereign in their governance, and that extends to their data and knowledge as well. It recognizes that Indigenous people are the ultimate authority on their data and knowledge and aims to redefine Indigenous peoples' relationships to research from being participants or subjects to being meaningful partners and co-researchers.

These ideas provide a challenge to dominant discourses in open data, questioning current approaches to data ownership, licensing, and use in ways that resonate beyond Indigenous contexts, drawing attention to the power and post-colonial dynamics within many data agendas. Growing networks for indigenous data sovereignty are working to shape open data principles to better respect the rights of Indigenous peoples.

6.3.1 Care Principles

The current movement toward open data and open science does not fully engage with Indigenous peoples' rights and interests. Existing principles within the open data movement (e.g. FAIR: findable, accessible, interoperable, reusable) primarily focus on characteristics of data that will facilitate increased data sharing among entities while ignoring power differentials and historical contexts. The emphasis on greater data sharing alone creates a tension for Indigenous peoples who are also asserting greater control over the application and use of Indigenous data and Indigenous knowledge for collective benefit.

This includes the right to create value from Indigenous data in ways that are grounded in Indigenous worldviews and realise opportunities within the knowledge economy. The *CARE Principles for Indigenous Data Governance* are people and purpose-oriented, reflecting the crucial role of data in advancing Indigenous innovation and self-determination. These principles complement the existing FAIR principles, encouraging open and other data movements to consider both people and purpose in their advocacy and pursuits.

The *CARE Principles for Indigenous Data Governance* guide the appropriate use and reuse of Indigenous data. This set of principles indicates the significant and crucial role of data in advancing Indigenous innovation and self-determination. CARE stands for (1) **collective** benefit, (2) **authority** to control, (3) **responsibility**, and (4) **ethics**. These can be described as follows:

1. ***Collective benefit***—Data ecosystems should be designed and function in ways that enable Indigenous peoples to derive benefit from the data.
2. ***Authority to control***—Indigenous peoples' rights and interests in Indigenous data must be recognized and their authority to control such data should be empowered. Indigenous data governance enables Indigenous peoples to determine how they are represented within data.
3. ***Responsibility***—Those working with Indigenous data have a responsibility to share how this data is used to support Indigenous peoples' self-determination and collective benefit.
4. ***Ethics***—Indigenous peoples' rights and wellbeing should be the primary concern at all stages of the data life cycle.

These *CARE Principles for Indigenous Data Governance* were drafted at the International Data Week and Research Data Alliance Plenary co-hosted event

"Indigenous Data Sovereignty Principles for the Governance of Indigenous Data Workshop," held on 8 November 2018 in Gaborone, Botswana.

6.3.2 Example: Indigenous Sovereignty in Canada

Canada has a long and complex history of navigating Indigenous-nonindigenous relations, going back to the beginning of colonial influence in what is now known as Canada. In this section, we will examine some of the history behind these legal relations, with the goal of better understanding the complexities and challenges which governments have tried—with limited success—to address.

Generally, issues relating to Indigenous rights in Canada can be divided into the following topical subsections:

- Resource rights
- Aboriginal title
- Right to self-governance
- Rights content
- Duty to consult
- Indigenous women's rights
- Indigenous rights extinguishment
- Métis rights
- Inuit rights

We will discuss these in the following sections, within the context of Canadian history and Indigenous relations within Canada.

6.3.2.1 Resource Rights

Historically, Indigenous peoples have had to prove their rights in Canadian courts. For resource rights other than Aboriginal title, the Supreme Court has held that Indigenous people must demonstrate that the right was integral to their distinctive societies and was exercised at the time of first contact with Europeans. What this means is that for practices such as fishing and hunting to be enshrined as rights, Indigenous peoples must prove that these activities were practised before the arrival of Europeans. The courts have seen commercial trade in furs and fish, for example, as the product of European contact rather than integral to Indigenous societies prior to contact. Fishing for food, community, or ceremonial purposes is, however, a protected right and may be exercised in a modern way with modern fishing equipment.

Indigenous peoples have used Sect. 35 of the *Constitution Act* to support their rights to resource activities, such as fishing. In the *Sparrow* case (1990)—the first decision by the Supreme Court to interpret Sect. 35—an Indigenous person fished contrary to the provisions of federal law. In his defence, he alleged that the right to fish was an immemorial right protected by treaty by virtue of Sect. 35. The Supreme Court upheld the right and set out a code of interpretation for Sect. 35. The court did not set limits on the types of rights that can be categorised as Indigenous rights and emphasised that the rights must be interpreted flexibly in a manner "sensitive to the aboriginal perspective." The court stated that Sect. 35 only protects rights that were not extinguished (i.e. surrendered) prior to the date the *Constitution Act, 1982*, came into effect.

Indigenous peoples have also defended their lands and rights to resources outside the courts. Protests against development companies and the government that seek to infringe on ancestral rights have demonstrated Indigenous resistance and the desire for consultation and open dialogue about matters that affect traditional lands and rights. Some well-known examples of such demonstrations include *Idle No More*, the *War in the Woods* (1984 to 1993), a protest led by the *Tla-o-qui-aht* and their allies against logging and deforestation in ancient forests, and protests against pipeline developments, such as the Mackenzie Valley and Keystone XL pipelines.

6.3.2.2 Aboriginal Title

There have been a few key court cases that have helped to define Aboriginal title. The *Calder* case (1973) recognized for the first time that Aboriginal title has a place in Canadian law. In the Delgamuukw case (1997), the Supreme Court ruled that claims to traditional lands had to show exclusive occupation of the territory by a defined Aboriginal society at the time the Crown asserted sovereignty over that territory. In the same case, the court ruled that the oral histories of Aboriginal peoples were to be accepted as evidence proving historic use and occupation. The *Tsilhqot'in* case (2014) further clarified the requirements for establishing Aboriginal title. The criteria for Aboriginal title are threefold: in short, an Aboriginal group must first prove occupation, and then must prove continuity and exclusivity of said occupation.

However, the court has not fully resolved all legal issues concerning Aboriginal title. Serious disputes have arisen over whether or not Aboriginal title carries with it the exclusive right to use and occupy lands. This is an issue in cases where the current occupation is not exclusively Indigenous people and where resource companies and other interests seek to carry on or expand their own uses of the same lands. Several court cases, including those involving the *Nuu-chah-nulth* in British Columbia, have already been launched over these issues. In most cases, the rulings ensure that proper administrative requirements are met, while permitting resource exploitation and development to continue in the overall public interest. The duty to consult was affirmed by the Supreme Court in the *Delgammuuk* case and is also a key part of the UN *Declaration on the Rights of Indigenous Peoples*.

6.3.2.3 Rights to Self-Government

Although Indigenous rights have yet to be given a comprehensive definition in law, most Indigenous peoples assert that they include the right to self-government. The Supreme Court has not directly addressed that issue. This was, however, a subject extensively studied by the Royal Commission on Aboriginal Peoples, which reported to the federal government in 1996. The Royal Commission proposed solutions for a new and better relationship between Indigenous peoples and the Canadian government, including recognition of the right of self-government, settlement of land claims, measures to eliminate inequalities between Indigenous and non-Indigenous peoples in Canada, and the creation of Indigenous justice systems.

One of the most well-known examples of self-government in Canada is the *Nisga'a Final Agreement,* signed after 25 years of negotiation following the *Calder* case in 1973. The content of the treaty and the ratification process were subjected to intense debate and were challenged in court. Upon Parliament's passage of the *Act* in 2000, the treaty became the first modern-day treaty in British Columbia and the 14th modern-day treaty in Canada to be negotiated from 1975 to 2000. The *Nisga'a Final Agreement* gave the First Nation the right to self-government within the 2019 km^2 in the Nass Valley to which the Nisga'a hold title. Since 1973, there have been 26 comprehensive land claims and four self-government agreements (as of 2015).

The *Nisga'a Final Agreement* was groundbreaking for the British Columbia treaty process because it achieved the aspirations for a negotiated settlement as expressed by the courts in the *Delgamuukw* case. Other First Nations in British Columbia continue negotiations on their claims. The Tsawwassen First Nation and the Maa-nulth First Nations finalised agreements in 2009 and 2011, respectively. As of July 2017, there were 58 ongoing comprehensive claims negotiations in British Columbia and another seven claims in the implementation process.

6.3.2.4 Rights Content

No Indigenous right, even though constitutionally protected, is absolute in Canadian law. Fishing rights, for example, are not exclusive in the sense that only Indigenous peoples can exercise them. Also, Indigenous rights are not immune to regulation by other governments. Additionally, Aboriginal title may give rise to an exclusive right to use and occupy lands, but that right may be infringed upon by the government for purposes such as economic development, power generation, or the protection of the environment or endangered species. However, non-Indigenous governments must justify infringement of Aboriginal rights or title on the basis of a legitimate government purpose and recognition of the constitutional protection of the rights being affected. There may also be a requirement for prior consultation with the Indigenous peoples concerned and compensation in some circumstances.

6.3.2.5 Duty to Consult

The duty to consult—and the issue of what levels of government are entitled or required to consult—has been further explored in two 2014 Supreme Court cases, *Grassy Narrows* and *Tsilhqot'in*. In Ontario, the *Grassy Narrows* case pushed forward the notion that provincial governments may also "take up" treaty lands for development, but in doing so, they also take on the federal government's responsibilities to consult with Indigenous peoples.

In the *Tsilhqot'in* case, the Supreme Court recognized the Aboriginal Title of the First Nation and the corresponding authority over 1750 km^2 of their traditional territory in the British Columbia interior. In taking an expansive view of Aboriginal title, the Supreme Court charted a new course relative to future resource development and the process of consulting with Indigenous groups in areas of Canada that have not been ceded by historic treaties. This suggests that the Crown in future must do more than fulfil a duty to consult. It must also either obtain consent or meet legal requirements to justify infringing on Indigenous rights.

6.3.2.6 Indigenous Women's Rights

During the 1970s and 1980s, Indigenous women launched several cases against the federal government concerning the legal and gender discrimination inherent in the *Indian Act*. Since 1869, Status Indian women who married non-status Indian men lost any treaty and Indigenous rights that they previously enjoyed. In 1985, Bill C-31 amended the *Indian Act* to remove the discrimination and bring the act into line with the *Canadian Charter of Rights and Freedoms*.

The 1985 amendment allows women who "married out" to apply for the restoration of their status and rights, and also allows their children to apply for registration as Status Indians. The *Act* no longer requires or allows women to follow their husbands into or out of status, and it allows women to pass status on to their children just as men always have.

However, while the amendment addressed much of the former discrimination against women, it also created some problems. By placing these women and often their children on First Nations band membership lists, the government stretched already limited lands and funds to serve more people, and this has at times incited resentment and backlash toward these "Bill C-31s" by First Nations members.

Further, the inclusion of a "second-generation cut-off" rule potentially means a great reduction in the number of people entitled to be registered as status Indians under the *Indian Act*. According to Bill C-31, there are two categories of Indian registration. The first, known as sub-sect. 6(1), applies when both parents are or were entitled to registration. The second, known as sub-sect. 6(2), applies when one parent is entitled to registration under 6(1.) Status cannot be transferred, however, if that one parent is registered under sub-sect. 6(2.) In short, after two generations of intermarriage with non-Status partners, children would no longer be eligible for status. Moreover, for a child to be registered, both the mother's and father's names must be included on the birth certificate. If the father's name is not included, he is assumed to be non-Status. In such situations, children born to women registered under sub-sect. 6(2) are not eligible for status. The amendment therefore significantly limits the ability to transfer status to one's children.

Bill C-31 also did away with the "double mother rule," which had conferred status on non-married individuals up to 21 years of age whose mother and paternal grandmother were non-Status. However, eliminating the effects of this rule created a new inequality that made it difficult to transmit status in certain cases.

In response to the British Columbia Court of Appeal ruling in *McIvor v. Canada* (2009), Bill C-3, passed in 2010, attempted to ensure parity of status for grandchildren of women who "married out" and those affected by the "double mother" rule.

6.3.2.7 Rights Extinguishment

The "extinguishment" of rights means the taking away or surrendering of rights. Historically, treaties or land claim settlements have served to extinguish Indigenous rights. All courts have recognized the power of Parliament to extinguish Aboriginal rights and title up to 1982, but this was never expressly done. Indigenous rights to hunt and fish, however, have been limited by constitutional amendment, federal legislation, and in some instances by provincial laws. In the 1990 *Sparrow* decision, the Supreme Court ruled that rights could be regulated if the regulation could be justified in the manner described above. In the *Delgamuukw* case, the Court did not rule out extinguishment after 1982, but made strong statements about consultation and compensation if rights are extinguished.

In the *Bear Island* case, a case appealed to the Supreme Court but dismissed in 1991, it was also held that delay in bringing a court action was sufficient to defeat a claim to Aboriginal title. This alone, if correct in law, would be enough to defeat almost every land claim that is brought to court. Moreover, the First Nations had signed a treaty in 1850, thereby extinguishing their rights. In the 1995 *Blueberry River* case, the Supreme Court applied a statutory limitation period to defeat part of a First Nation's claim with respect to a surrender of reserve lands and mineral rights.

Historically, federal laws have also worked to deny the rights of Indigenous peoples. The *Indian Act* has taken away basic rights over time, such as the right to hold potlatches, dance, and practice Indigenous religions. Willing or forcible enfranchisement (the processes by which Indigenous peoples lost their status under the *Indian Act*) extinguished Indigenous rights in exchange for others as Canadian citizens. Other federal legislation has also worked to assimilate Indigenous peoples and therefore deny them their rights, such as residential schools and the pass system (a policy in which Indigenous peoples who wished to leave their reserve, even temporarily, had to acquire a pass from an Indian agent before leaving).

6.3.2.8 Metis Rights

Métis rights have largely been defined and clarified by the courts. A particularly important case is *R. v. Powley* (2003), which affirmed the Métis ancestral right to hunt for sustenance. *Powley* was the first case in which the Supreme Court affirmed the existence of Métis rights. The case also established a test to determine Métis rights under Sect. 35 of the *Constitution Act, 1982*. The *Powley* test has 10 criteria that determine Métis identity and if a Métis community has an existing right to an activity, such as hunting.

The identity part of the test consists of three criteria: to be considered Métis, individuals must identify as Métis, be a member of a modern Métis community, and have ties to a historic Métis community. This last requirement also asks that individuals prove another set of criteria: that their mixed ancestry group formed a "distinctive" collective social identity, that they lived together in the same geographic area, and that they shared a common way of life.

The *Powley* case therefore confirmed the definition of Métis as one that is specific to a distinct community of peoples rather than anyone who has mixed Indigenous and European heritage. For those who consider themselves Métis but do not fit this description—namely those living outside the territory as defined by the Métis National Council, including Alberta, Manitoba, Saskatchewan, and parts of Ontario, British Columbia, and the Northwest Territories– they feel the *Powley* decision denies their Indigenous identity. However, for those who can trace their lineage to specific, historic Métis communities, the *Powley* case validates their long-standing argument that they are a distinct Indigenous community.

The *Daniels* court case (2016) is also significant to Métis rights. On 14 April 2016, the Supreme Court ruled in the *Daniels* decision unanimously that the legal definition of "Indian"—as laid out in the *Constitution*—now includes the Métis and non-status Indians. This ruling will facilitate possible negotiations over traditional land rights, access to education and health programs, and other government services. This ruling did not, however, grant Indian status to any Métis or non-status Indian people.

6.3.2.9 Inuit Rights

The Inuit fall under the category of "Indian" in the *Constitution Act, 1982*, and are therefore also protected by Sect. 35. However, the Inuit have never been subject to the *Indian Act* and were largely ignored by the federal government until 1939, when a court decision ruled that they were a federal responsibility, though still not subject to the *Indian Act*. Policies of assimilation followed, including forced relocations into sedentary communities and the introduction of disk numbers for administrative purposes.

Various treaties and land claims have confirmed Inuit rights to land title in northern Canada, including the *James Bay and Northern Québec Agreement*

(1975), the *Western Arctic (Inuvialuit) Claims Settlement Act* (1984), the *Nunavut Land Claims Agreement Act* (1993) and the *Labrador Inuit Land Claims Agreement* (2005.) Together, these four regions cover about 40 percent of Canada's land mass.

An important court case concerning Inuit rights is the *Clyde River* case (2017.) The Inuit living in and around Clyde River, Nunavut, vehemently opposed the plans of the National Energy Board (NEB) to conduct seismic testing for oil and gas deposits near their community since it was first proposed in 2011. Taking their case to the Supreme Court, the Clyde River Inuit emerged victorious: the judges unanimously decided that the NEB failed to properly consult the Inuit about their plans and did not adequately assess the effect that seismic testing would have on the rights of Indigenous peoples. Quashing the NEB's approval, the Supreme Court put an end to the seismic testing. While the court did not create guidelines for how to consult with an Indigenous community, this case highlights the importance of consultation.

On the international stage, Inuit peoples have had their rights to Arctic lands and waters affirmed by such declarations as the *United Nations Declaration on the Rights of Indigenous Peoples.*

There is still much work to be done in Canada with regard to indigenous relationships, and the work on this is only truly just beginning. The work done and the recommendations made by the Truth and Reconciliation Committee, which held hearings across Canada, will be an important step in moving forward to address these historical inequalities and to honour the commitments made in the nation-to-nation treaties between the colonising groups and indigenous nations over Canada's modern history.

6.4 Data Centres and Data Mines

Data mining is the process of finding useful new correlations, patterns, and trends by transferring through a high amount of data saved in repositories, using pattern recognition technologies including statistical and mathematical techniques. It is the analysis of factual datasets to discover unsuspected relationships and to summarise the records in novel methods that are both logical and helpful to the data owner.

6.4.1 Functions

There are various concepts and functions involved in data mining, including classification, predictions, association rules and recommendation systems, and data reduction.

6.4.1.1 Classification

Classification is the procedure of discovering a model that represents and distinguishes data classes or concepts, with the objective of being able to use the model to predict the class of objects whose class label is anonymous. The derived model is based on the analysis of a group of training records, that is, data objects whose class label is familiar.

6.4.1.2 Predictions

Prediction is the same as classification, except that for prediction, the results are misrepresented in the future.

Examples of prediction functions in business and research include:

- It can predict the value of a stock three months into the future.
- It can predict the percentage increase in traffic deaths next year if the speed limit is raised.
- It can predict the winner of this fall's baseball World Series, based on a correspondence of team statistics.
- It can predict whether a definite molecule in drug discovery will lead to a cost-effective new drug for a pharmaceutical company.

6.4.1.3 Association Rules and Recommendation Systems

Association rules, or affinity analysis, are designed to find such general association patterns between items in large databases. The rules can be used in several methods. For example, grocery stores can use such information for product placement.

They can use the rules for weekly promotional offers or for bundling products. Association rules derived from a hospital database on patients' symptoms during consecutive hospitalizations can help find "which symptom is followed by what other symptom" to help predict future symptoms for returning patients.

6.4.1.4 Data Reduction

Data mining is used to select data in a huge number of databases. When data analysis and mining are completed on a huge number of records, then it takes a very long time to process, which makes it impossible and infeasible.

It can reduce the processing time for data analysis. **Data reduction techniques** are used to obtain a reduced representation of the dataset that is much smaller in volume by maintaining the integrity of the original data. By reducing the data, the efficiency of the data mining process is improved, which produces the same analytical results.

Data reduction aims to define it more compactly. When the data size is smaller, it is easier to use mature and computationally high-cost algorithms. The reduction of the data may be in terms of the number of rows (records) or terms of the number of columns (dimensions).

6.4.2 Social Impact

Data mining is the process of finding useful new correlations, patterns, and trends by transferring through a high amount of data saved in repositories, using pattern recognition technologies including statistical and mathematical techniques. It is the analysis of factual datasets to discover unsuspected relationships and to summarise the records in novel methods that are both logical and helpful to the data owner.

Data mining systems are designed to promote the identification and classification of individuals into different groups or segments. From the perspective of the commercial firm and possibly for the industry as a whole, it can interpret the use of data mining as a discriminatory technology in the rational search of profits. There are various social implications of data mining, including privacy, profiling, and unauthorised use.

6.4.2.1 Privacy

Privacy is a loaded issue. In current years, privacy concerns have taken on a more important role in American society as merchants, insurance companies, and government agencies amass warehouses including personal records.

The concerns that people have over the group of this data will generally extend to some analytic capabilities used to the data. Users of data mining should start thinking about how their use of this technology will be impacted by legal problems associated with privacy.

6.4.2.2 Profiling

Data mining and profiling is a developing field that attempts to organise, understand, analyse, reason, and use the explosion of data in this information age. The process involves using algorithms and experience to extract designs or anomalies that are very complex, difficult, or time-consuming to recognize.

The founder of Microsoft's Exploration Team used complex data mining algorithms to solve an issue that had haunted astronomers for some years. The problem of reviewing, describing, and categorising two billion sky objects recorded over three decades is that the algorithm extracted the relevant design to allocate the sky objects like stars or galaxies. The algorithms were able to extract the features that represented sky objects such as stars or galaxies. This developing field of data mining and profiling has several frontiers where it can be used.

6.4.2.3 Unauthorised Use

Trends obtained through data mining designed to be used for marketing goals or some other ethical goals can be misused. Unethical businesses or people can use the data obtained through data mining to take advantage of vulnerable people or discriminate against a specific group of people. Furthermore, the data mining technique is not 100 percent accurate; thus, mistakes do appear which can have serious results.

6.4.3 Environmental Impacts

For years, bitcoin critics have maligned the world's biggest cryptocurrency for polluting the planet. But new data from Cambridge University shows that the geography of mining has drastically changed over the last six months, which will improve bitcoin's carbon footprint.[48]

China's big crypto crackdown this spring set off a chain reaction in the mining world. For one, it took half the world's bitcoin miners offline practically overnight.[49] Fewer people mining has meant less machines running and less power being consumed overall, which slashed bitcoin's environmental impact. Beijing's new crypto rules also permanently took a lot of older and more inefficient gear offline. And crucially, China shutting its doors to crypto mining has set off a massive migration. Miners are now heading to the cheapest sources of energy on the planet, which more often than not are renewable.

China has long been the mecca of the crypto mining world, accounting for nearly three-quarters of all bitcoin miners at its peak, according to the Cambridge Centre for Alternative Finance. But after Beijing decided to expel its miners in May, more than 50% of the hash rate—the collective computing power of miners worldwide—dropped off the network.[50,51]

Today, bitcoin draws roughly 70 terawatt hours of energy per year, or 0.33% of the world's total electricity production.[52] That is almost half of what it was in May and is roughly equivalent to the annual energy draw of countries like Bangladesh and Chile.

[48] https://cbeci.org/mining_map

[49] https://www.blockchain.com/charts/hash-rate

[50] http://www.gov.cn/xinwen/2021-05/21/content_5610192.htm

[51] https://www.blockchain.com/charts/hash-rate

[52] https://cbeci.org/cbeci/comparisons

The exodus from China also means that a lot of older mining equipment that was probably long-past due for retirement will never be turned back on. But not all of China's miners are going dark. Many have begun to patronise elsewhere, gravitating to the world's cheapest sources of power.

Because miners at scale compete in a low-margin industry, where their only variable cost is typically energy, they are incentivized to migrate to the world's cheapest sources of power. In effect, they need to constantly reduce their electricity costs, which is their number one expense, in order to be competitive.

The data shows that a whole lot of these miners are headed for cheaper pastures in the United States. The United States has fast become the new hotspot for the world's global crypto miners. In the last six months, the country has jumped from fifth to second place and now accounts for nearly 17% of all global bitcoin miners. Although China was still solidly in first place as of April, with a 46% share, America's share of the market is likely a lot higher now since the Chinese government booted miners in May. US-based bitcoin mining operators have seen a huge uptick in business. Indeed, retail hardware and hosting sales have increased nearly 300% since mid-June.

Energy consumption is not equivalent to carbon emissions. While it is relatively easy to determine the amount of energy that is consumed by the bitcoin network, it is much harder to determine its carbon footprint. An accurate read of bitcoin's carbon emissions would require exact knowledge of the energy mix used to generate electricity used by each bitcoin mining operation. One unit of hydropower, for example, does not have the same environmental impact as the equivalent amount of power sourced from coal. And China's bitcoin mining operations were known for both.

But overall, the market is pushing North American energy sources to get greener. Each year, investment bank Lazard releases a breakdown of energy costs by source. Its 2020 report shows that many of the most common renewable energy sources are either equal to or less expensive than conventional energy sources like coal and gas.[53] And the cost of renewable power keeps going down.[54]

It has been indicated that most miners new to North America will be powered by renewables, or gas offset by renewable energy credits, with one party estimating that bitcoin mining in the United States is more than 50% powered by renewables. Miners migrating to North America are also preparing for a future in which their energy usage is questioned by putative investors—and possibly regulated.

Many Chinese mining clients are aware of the political and normative winds in North America and want to hedge themselves against regulatory risks in the future by establishing new facilities in primarily renewable-powered locations. Indeed, the largest of them are also looking at the potential of going public or are looking for investors to help them grow, as they realise that public markets nowadays have no

[53] https://www.lazard.com/perspective/levelized-cost-of-energy-and-levelized-cost-of-storage-2020/

[54] https://ourworldindata.org/cheap-renewables-growth

appetite for proof of work mining that is powered by non-renewable [energy sources].

Not all miners, however, are headed to renewable destinations. Kazakhstan is now just behind the U.S. in terms of its share of the global bitcoin mining market, with about 8% of all crypto mining. It's home to coal mines that provide a cheap and abundant supply of energy—but also ample carbon dioxide emissions.

However, several mining experts have indicated that they think that Kazakhstan, which neighbours China, is just a temporary stopover on a longer migration west, with large miners going there in the short-term with older-generation equipment. As older-generation machines reach the end of their service lives, those companies will likely deploy new machines into more stable and energy efficient and renewable jurisdictions. Another potential damper on Kazakhstan's popularity is a law newly signed by the president that will introduce extra taxes for crypto miners starting in 2022, which could significantly change the incentives for people to deploy capital in Kazakhstan.[55]

6.4.4 Challenges

Challenges for data mining include efficiency and scalability of data mining algorithms; usefulness, certainty, and expressiveness of data mining results; expression of various kinds of data mining results; interactive mining knowledge at multiple abstraction levels; and mining information from different sources of data.

6.4.4.1 Efficiency and Scalability of Algorithms

It can effectively extract data from a large amount of data in databases. The knowledge discovery algorithms should be efficient and scalable to huge databases. Specifically, the running time of a data mining algorithm should be predictable and acceptable in huge databases. Algorithms with exponential or even channel-order polynomial complexity will not be of efficient use.

6.4.4.2 Usefulness, Certainty, and Expressiveness of Results

The identified knowledge should exactly portray the contents of the database and be beneficial for specific applications. Imperfectness must be defined by measures of uncertainty, in the form of approximate rules or quantitative rules.

Noise and exceptional data must be managed elegantly in data mining systems. This also stimulates a systematic study of measuring the quality of the discovered

[55] https://kursiv.kz/news/obschestvo/2021-06/tokaev-podpisal-chetyre-zakona

knowledge, such as interestingness and reliability, through the development of statistical, analytical, and simulative models and tools.

6.4.4.3 Expression of Various Kinds of Results

Several kinds of knowledge can be discovered from a huge amount of data. It can also like to examine discovered knowledge from multiple views and display them in different forms.

This required us to define both the data mining requests and the discovered knowledge in high-level languages or graphical user interfaces so that the data mining task can be defined by non-experts and the discovered knowledge can be understandable and precisely available by users. This also needed the discovery system to select expressive knowledge representation techniques.

6.4.4.4 Interactive Mining Knowledge at Multiple Levels

Because it is complex to predict what exactly can be discovered from a database, a high-level data mining query must be considered as a probe that can disclose some interesting traces for further exploration.

Interactive discovery must be encouraged, which enables a user to interactively refine a data mining request, dynamically change data focusing, progressively deepen a data mining process, and flexibly view the information and data mining results at several abstraction levels and from multiple angles.

6.4.4.5 Mining Information from Different Sources of Data

The broadly available local and wide-area computer networks, such as the Internet, can connect various sources of data and form huge distributed, heterogeneous databases. Mining knowledge from multiple sources of formatted or unformatted information with diverse data semantics poses a new requirement for data mining.

Otherwise, data mining can help disclose the high-level data regularities in heterogeneous databases which can barely be discovered by simple query systems. Furthermore, the huge size of the database, the broad distribution of data, and the computational complexity of several data mining methods motivate the advancement of parallel and distributed data mining algorithms.

6.5 Digital Self-Determination

Digital self-determination is a multidisciplinary concept derived from the legal concept of self-determination and applied to the digital sphere, to address the unique challenges to individual and collective agency and autonomy arising with the increasing digitalization of many aspects of society and daily life. Broadly speaking, the term describes the attempt to comprehensively project the pattern of human self-determination (as first explored in disciplines like philosophy and psychology, and in the law) into the digital age.

The concept has been included in an official document for the first time by ARCEP, the French Telecoms Regulator, in a section of its *2021 Report on the State of the Internet*,[56] exploring the work on "Network Self-determination" conducted by Professor Luca Belli.[57]

The increasing digitization of most aspects of society poses new challenges for the concept and realisation of self-determination.[58] While the digital sphere has ushered in innovation and opened up new opportunities for self-expression and communication for individuals across the globe, its reach and benefits have not been evenly distributed, oftentimes deepening existing inequalities and power structures, commonly referred to as a digital divide. Moreover, the digital transformation has enabled, oftentimes unbeknownst to individuals, the mass collection, analysis, and harvesting of personal data by private companies and governments to infer individuals' information and preferences (e.g. by tracking browsing and shopping history), influence opinions and behaviour (e.g. through filter bubbles and targeted advertisements), and/or to make decisions about them (e.g. approving or not a loan or employment application), thus posing new threats to individuals' privacy and autonomy.[59]

Although the definition of digital self-determination is still evolving, the term has been used to address humans' capacity (or lack thereof) to exercise self-determination in their existence in and usage of digital media, spaces, networks, and technologies, with the protection of the potential for human flourishing in the digital world as one of the chief concerns.[60]

---

[56] *The State of Internet in France* (PDF). ARCEP. 2021. https://en.arcep.fr/uploads/tx_gspublication/report-state-internet-2021-edition-july2021.pdf

[57] Belli, Luca (2018). "Network self-determination: When building the Internet becomes a right". *IETF Journal*. https://www.ietfjournal.org/network-self-determination-when-building-the-internet-becomes-a-right/

[58] Remolina, Nydia and Findlay, Mark James, The Paths to Digital Self-Determination – A Foundational Theoretical Framework (April 22, 2021). SMU Centre for AI & Data Governance Research Paper No. 03/2021.

[59] Center for Humane Technology, "Ledger of Harms", https://ledger.humanetech.com/

[60] Berkman Klein Center for Internet and Society at Harvard University, Digital Self-Determination call for participants (2021) <https://cyber.harvard.edu/story/2021-02/call-participants

Starting in the 2010s, a few multidisciplinary and cross-sectoral initiatives around the world have been working on developing a theoretical framework for the concept of digital self-determination.

6.5.1 Theories of Self-Determination

Historically, the concept of self-determination relates to concepts of subjectivity, dignity, and autonomy in classic central-European philosophy and is derived from Immanuel Kant's conception of freedom. Self-determination presupposes that human beings are entities capable of reason and responsibility for their own rationally chosen and justified actions (autonomy), and ought to be treated accordingly. In formulating his categorical imperative (kategorischer Imperativ), Kant suggested that humans, as a condition of their autonomy, must never be treated as a means to an end but as an end in itself. The pattern of self-determination similarly aims at enabling autonomous human beings to create, choose, and pursue their own identity, actions, and life choices without undue interference.

In psychology, the concept of self-determination is closely related to self-regulation and intrinsic motivation, i.e. engaging in a behaviour or activity because it is inherently rewarding to do so, as opposed to being driven by external motivations or pressures, like monetary incentives, status, or fear. In this context, self-determination and intrinsic motivation are linked to feeling in control of one's choices and behaviour and are considered necessary for psychological well-being. Self-determination theory (SDT), first introduced by psychologists Richard Ryan and Eduard Deci in the 1980s,[61] and further developed through the 1990s and 2000s, has been largely influential in shaping the concept of self-determination in the field of psychology.[62]

Ryan and Deci's self-determination theory proposed that individuals' motivated behaviour is characterised by three basic and universal needs: autonomy, competence, and relatedness.[63] Autonomy refers here to the need to feel free to decide one's course of action. Competence refers to the need to have the capacity and skills to undertake and complete motivated behaviour in an effective manner. Finally, relatedness refers to the need to experience warm and caring social relationships and feel connected to others. According to self-determination theory, all three needs must be fulfilled for optimal functioning and psychological well-being. However, other psychologists like Barry Schwartz have argued that if self-determination is

---

[61] Deci, E. L., & Ryan, R. M. (1985). Intrinsic motivation and self-determination in human behavior. New York: Plenum.

[62] Deci, E. L., & Ryan, R. M. (1980). Self-determination theory: When the mind mediates behavior. The Journal of Mind and Behavior, 33–43.

[63] Ryan, R. M.; Deci, E. L. (2000). "Self-determination theory and the facilitation of intrinsic motivation, social development, and well-being". American Psychologist. 55 (1): 68–78.

taken to extremes, freedom of choice can turn into the "tyranny of choice".[64] In this view, having too much autonomy and too many choices over our course of action can be perceived as overwhelming, make our decisions more difficult, and ultimately lead to psychological distress rather than wellbeing.

In international law, the right of a people to self-determination is commonly recognized as a ius cogens rule. Here, self-determination denotes that a people, based on respect for the principle of equal rights and fair equality of opportunity, have the right to freely choose their sovereignty, international political status, and economic, social, and cultural development with no interference. In the framework of the United Nations, fundamental rights like self-determination are mainly defined in the *Universal Declaration of Human Rights*, the *International Covenant on Civil and Political Rights*, and the *International Covenant on Economic, Social, and Cultural Rights*.

6.5.1.1 Example: Self Determination in Germany

The concept of informational self-determination (informational Selbstbestimmung), considered as a modern fundamental right which protects against unjustified data processing, has featured prominently in the German Federal Constitutional Court's (Bundesverfassungsgericht) jurisprudence and might be the most direct precursor and inspiration to the concept of digital self-determination.

In 1983, the Bundesverfassungsgericht ruled that "in the context of modern data processing, the general right of personality under Article 2.1 in conjunction with Article 1.1 of the *Basic Law* encompasses the protection of the individual against unlimited collection, storage, use, and sharing of personal data. The fundamental right guarantees the authority conferred on the individual to, in principle, decide themselves on the disclosure and use of their personal data."[65]

Philosophically, the right to informational self-determination is deeply rooted in the Bundesverfassungsgericht's understanding of inviolable human dignity as a prohibition of human objectification.[66] This understanding refers back to the late eighteenth-century German philosophy of the Enlightenment. The Volkszählungsurteil was inspired by the concern that modern data processing technology could lead to a "registration and cataloguing of one's personality in a manner that is incompatible with human dignity".[67]

In this view, human beings, due to their inviolable dignity, may never be treated like depersonalised and objectified resources that can be harvested for data. Instead,

[64] Schwartz, B. (2000). Self-determination: The tyranny of freedom. American psychologist, 55(1), 79.

[65] Article 1.1 and 2.1 of German *Basic Law*.

[66] Article 1 of the Grundgesetz.

[67] Volkszählungsurteil, headnote 4.

humans, due to their capacity for autonomy, are self-determined agents possessing a significant degree of control over their informational images.

6.5.2 Elements of Digital Self-Determination

Different sectors of society, ranging from legislators and policymakers to public organisations and scholars, to activists and members of the civil society, have called for digital infrastructure, tools, and systems that protect and promote individuals' self-determination, including equal and free access, human-centred design, better privacy protections, and control over data. These elements are closely connected and complement one another.

For example, equal access to digital infrastructure can enable the representation of diverse viewpoints and participatory governance in the digital sphere, and decentralised systems might be necessary to ensure individuals' control over their data.

6.5.2.1 Access

Bridging the various forms of existing digital divides and providing equitable and fair access to digital technologies and the Internet has been proposed as crucial to ensure that all individuals are able to benefit from the digital age, including access to information, services, and advancement opportunities.[68,69]

In this sense, the concept of digital self-determination overlaps with the concept of "network self-determination" as it emphases that groups of unconnected and scarcely connected individuals can regain control over digital infrastructures, by building them and shaping the governance framework that will organise them as a common good.[70,71] As such, Belli stresses that network self-determination leads to several positive externalities for the affected communities, preserving the Internet as an open, distributed, interoperable, and generative network of networks.[72]

---

[68]Digital Divide, World Wide Web Foundation, https://webfoundation.org/news/tag/digital-divide/

[69]Closing the Digital Divide, Common Sense Media, https://www.commonsensemedia.org/kids-action/publications/closing-the-k-12-digital-divide-in-the-age-of-distance-learning

[70]Belli, Luca (2019). "La emergencia de las redes comunitarias y del principio de autodeterminación de red: un ejemplo de gobernanza de Internet". *Revista de privacidad y derecho digital.* **4** (13): 77–112.

[71]Belli, Luca (2018). "Community Networks: Bridging Digital Divides through the Enjoyment of Network Self-determination". *The Community Network Manual: how to build the Internet yourself* (PDF). FGV Direito Rio. pp. 23–41.

[72]Belli, Luca (2018). "Network self-determination: When building the Internet becomes a right". *IETF Journal.*

6.5.2.2 Literacy

Digital literacy and media literacy have been proposed as necessary for individuals to acquire the knowledge and skills to use digital tools as well as to critically assess the content they encounter online, create their own content, and understand the features and implications of the digital technology used on them as well as the technology they consciously and willingly engage with.[73] In addition to basic digital navigation skills and critical consumption of information, definitions of digital literacy have been extended to include an awareness of existing alternatives to the digital platforms and services used, understanding how personal data is handled, awareness of rights and existing legal protections, and measures to independently protect one's security and privacy online. For example, the adoption of obfuscation techniques as a way of evading and protesting digital surveillance.[74]

6.5.2.3 Design

Scholars have coined the term attention economy to refer to the treatment of human attention as a scarce commodity in the context of ever-increasing amounts of information and products. In this view, the increasing competition for users' limited attention, especially when relying on advertising revenue models, creates a pressing goal for digital platforms to get as many people as possible to spend as much time and attention as possible using their product or service. In their quest for users' scarce attention, these platforms would be incentivized to exploit users' cognitive and emotional weaknesses, for example, via constant notifications, dark patterns, forced multitasking, social comparison, and incendiary content. Advocates of human-centred design in technology (or humane technology) propose that technology should refrain from such "brain-hacking" practices, and instead should support users' agency over their time and attention as well as their overall wellbeing.[75]

6.5.2.4 Representation

Eli Pariser—an Internet activist, author, and entrepreneur—coined the term filter bubble to refer to the reduced availability of divergent opinions and realities that we encounter online as a consequence of personalization algorithms like personalised

[73] "What is digital literacy?", Common Sense Media, https://www.commonsensemedia.org/news-and-media-literacy/what-is-digital-literacy

[74] Brunton, F., & Nissenbaum, H. (2015). Obfuscation: A User's Guide for Privacy and Protest. MIT Press https://mitpress.mit.edu/books/obfuscation

[75] Center for Humane Technology, "How Social Media Hacks Our Brains", https://www.humanetech.com/brain-science

search and recommendation systems.[76] Filter bubbles have been suggested to facilitate a warped understanding of others' points of view and the world. Ensuring a wide representation of diverse realities on digital platforms could be a way of increasing exposure to conflicting viewpoints and avoiding intellectual isolation into informational bubbles.

6.5.2.5 Regulation

Scholar Shoshana Zuboff—Professor Emerita at Harvard Business School—coined the term surveillance capitalism to refer to the private sector's commodification of users' personal data for profit, leading to increased vulnerability to surveillance and exploitation.[77] Surveillance capitalism relies on centralised data management models wherein private companies retain ownership and control over the users' data.[78] To guard against the challenges to individuals' privacy and self-determination, various alternative data governance models have been recently proposed around the world, including trusts,[79] commons,[80] cooperatives,[81] collaboratives,[82] fiduciaries,[83] and "pods".[84] These models have some overlap and share a common mission to give more control to individuals over their data and thus address the current power imbalances between data holders and data subjects.

[76] Pariser, Eli (2011-05-12). The Filter Bubble: How the New Personalized Web Is Changing What We Read and How We Think. Penguin.

[77] Shoshana Zuboff an American author, Harvard professor, social psychologist, philosopher, and scholar.

[78] Kavenna, Joanna (October 4, 2019). "Shoshana Zuboff: 'Surveillance capitalism is an assault on human autonomy'". *The Guardian*.

[79] "What is a data trust?", Data Futures Lab, Mozilla Foundation,https://foundation.mozilla.org/en/data-futures-lab/data-for-empowerment/shifting-power-through-data-governance/#what-is-a-data-trust

[80] "What is a data common?", Data Futures Lab, Mozilla Foundation, https://foundation.mozilla.org/en/data-futures-lab/data-for-empowerment/shifting-power-through-data-governance/#what-is-a-data-commons

[81] "What is a data cooperative?", Data Futures Lab, Mozilla Foundation, https://foundation.mozilla.org/en/data-futures-lab/data-for-empowerment/shifting-power-through-data-governance/#what-is-a-data-cooperative

[82] "What is a data collaborative?", Data Futures Lab, Mozilla Foundation, https://foundation.mozilla.org/en/data-futures-lab/data-for-empowerment/shifting-power-through-data-governance/#what-is-a-data-collaborative

[83] "What is a data fiduciary?", Data Futures Lab, Mozilla Foundation, https://foundation.mozilla.org/en/data-futures-lab/data-for-empowerment/shifting-power-through-data-governance/#what-is-a-data-fiduciary

[84] Solid Project, https://solidproject.org/

6.6 Artificial Intelligence

> Every aspect of learning or any other feature of intelligence can be so precisely described that a machine can be made to simulate it.[85]

Artificial intelligence (AI) is intelligence demonstrated by machines, as opposed to the natural intelligence displayed by animals including humans.[86] AI research has been defined as the field of study of intelligent agents, which refers to any system that perceives its environment and takes actions that maximise its chance of achieving its goals.[87]

The field of artificial intelligence was founded on the assumption that human intelligence "can be so precisely described that a machine can be made to simulate it". The term "artificial intelligence" has previously been used to describe machines that mimic and display "human" cognitive skills that are associated with the human mind, such as "learning" and "problem-solving". This definition has since been rejected by major AI researchers who now describe AI in terms of rationality and acting rationally, which does not limit how intelligence can be articulated.[88]

Artificial intelligence was founded as an academic discipline in 1956 and in the years since has experienced several waves of optimism,[89] followed by disappointment and the loss of funding,[90] followed by new approaches, success, and renewed funding.[91] AI research has tried and discarded many different approaches since its founding, including simulating the brain, modelling human problem solving, formal logic, large databases of knowledge, and imitating animal behaviour. In the first decades of the twenty-first century, highly mathematical-statistical machine learning has dominated the field, and this technique has proved highly successful, helping to solve many challenging problems throughout industry and academia.[92]

The various subfields of AI research are centred around goals and the use of particular tools. The traditional goals of AI research include reasoning,[93] knowledge representation, planning, learning, natural language processing, perception, and the

---

[85] McCarthy, John; Minsky, Marvin; Rochester, Nathan; Shannon, Claude (1955). "A Proposal for the Dartmouth Summer Research Project on Artificial Intelligence".

[86] Russell, Stuart J.; Norvig, Peter (2003), *Artificial Intelligence: A Modern Approach* (second ed.), Upper Saddle River, New Jersey: Prentice Hall.

[87] Poole, David; Mackworth, Alan; Goebel, Randy (1998). *Computational Intelligence: A Logical Approach*. New York: Oxford University Press.

[88] Russell, Stuart J.; Norvig, Peter (2009). *Artificial Intelligence: A Modern Approach* (third ed.). Upper Saddle River, New Jersey: Prentice Hall.

[89] Crevier, Daniel (1993). *AI: The Tumultuous Search for Artificial Intelligence*. New York, NY: BasicBooks.

[90] A phenomenon known as the "AI winter".

[91] Clark, Jack (2015b). "Why 2015 Was a Breakthrough Year in Artificial Intelligence". *Bloomberg.com*

[92] Kurzweil, Ray (2005). *The Singularity is Near*. Penguin Books.

[93] Or "automated reasoning".

ability to move and manipulate objects.[94] General intelligence (the ability to solve an arbitrary problem) is among the field's long-term goals.[95] To solve these problems, AI researchers have adapted and integrated a wide range of problem-solving techniques—including search and mathematical optimization, formal logic, artificial neural networks, and methods based on statistics, probability, and economics. AI also draws upon computer science, psychology, linguistics, philosophy, and many other fields.

As machines become increasingly capable, tasks considered to require "intelligence" are often removed from the definition of AI, a phenomenon known as the AI effect.[96] For instance, optical character recognition is frequently excluded from things considered to be AI, having become a routine technology.[97] That said, some of the current applications being explored for artificial intelligence include:

- Advanced web search engines (e.g. Google)
- Recommendation systems (e.g. the types used by YouTube, Amazon, and Netflix)
- Understanding human speech (e.g. Siri and Alexa)
- Self-driving cars (e.g. Tesla)
- Automated decision-making
- Competing in high-level strategic games

6.6.1 Artificial Intelligence and Intellectual Property

Intellectual property is a category of property that includes intangible creations of the human intellect. There are many types of intellectual property, and some countries recognize more than others. The best-known types are copyrights, patents, trade secrets, industrial designs, and trademarks.

We can define each of these, based on definitions from the Canadian Intellectual Property Office, as:

6.6.1.1 Copyright

Copyright is the exclusive legal right to produce, reproduce, publish, or perform an original literary, artistic, dramatic, or musical work, or a substantial part of that work, in any form. It includes the right to perform the work or any substantial part of it or,

---

[94] Poole, David; Mackworth, Alan; Goebel, Randy (1998). *Computational Intelligence: A Logical Approach*. New York: Oxford University Press.

[95] Pennachin, C.; Goertzel, B. (2007). "Contemporary Approaches to Artificial General Intelligence". *Artificial General Intelligence*. Cognitive Technologies. Berlin, Heidelberg: Springer.

[96] McCorduck, Pamela (2004), *Machines Who Think* (second ed.), Natick, MA: A. K. Peters, Ltd.

[97] Schank, Roger C. (1991). "Where's the AI". *AI magazine*. Vol. 12, no. 4.

in the case of a lecture, to deliver it. If the work is unpublished, copyright includes the right to publish the work or any substantial part of it. Copyright provides protection for literary, artistic, dramatic, or musical works and other subject matter known as the performer's performances, sound recordings, and communication signals.

The creator is usually the copyright owner. However, an employer—for example, a film studio—may have copyright on works created by employees unless there is an agreement in place stating otherwise. When you own the copyright in a work, you control how it is used to protect its value. Others who want to use the work must either buy or otherwise get your permission.

Generally, an original work is automatically protected by copyright the moment you create it. By registering your copyright, you receive a certificate that can be used in court as evidence that you own it. Your copyright exists during your lifetime and for 50 years following your death. After that, the work is in the public domain, and anyone can use it. This is true for most works, but there are exceptions.

6.6.1.2 Patent

Patents cover new inventions or any new and useful improvement to an existing invention. A patent is granted to you by the government. It gives you the right to exclude others from making, using, or selling your invention. When you apply for a patent, you must provide a full description of the invention so that others can benefit from this advance in technology and knowledge. Patents are a way for people to share cutting-edge information. Each patent document describes a new aspect of a technology in clear and specific terms and is available for anyone to read. This makes it a vital resource for businesses, researchers, academics, and others who need to keep up with developments in their fields.

Through a patent, the government gives you, the inventor, the right to stop others from making, using, or selling your invention from the day the patent is granted to a maximum of 20 years after the day on which you filed your patent application. Patent protection applies in the country or region that issues the patent. Patents can have a great deal of value. You can sell them, licence them, or use them as assets to attract funding from investors.

For example, a Canadian patent applies within Canada for 20 years from the date you file the application, and the patent application is available to the public 18 months after you file it.[98] The rights granted by a Canadian patent extend throughout Canada, but not to other countries. You must apply for patent rights in other countries separately. Likewise, foreign patents do not protect an invention in Canada.

[98] http://www.ic.gc.ca/eic/siTe/cipointernet-internetopic.nsf/eng/wr03987.html

6.6.1.3 Trade Secret

Trade secrets include any business information that has commercial value derived from its secrecy. Trade secrets can be very valuable to you, whether you have developed new technology, designed original products, created the perfect recipe, or amassed a gold mine of customer data. Trade secrets include various assets such as sales methods, distribution methods, customer profiles, client lists, supplier lists, product ingredients and formulas, etc. One of the most famous trade secrets is the Coca Cola formula—a well-guarded secret for over 100 years. The commercial value of the formula is why the company goes to great lengths to keep it secret.

In Canada, for example, unlike for some other types of intellectual property, there is no formal process for protecting a trade secret. Rather, there is a body of established common law precedent on which the court relies in order to determine whether something is a trade secret. The protection of a trade secret requires the following, at a minimum[99]:

- That the information has commercial value
- That the information is secret
- That the information has been subject to reasonable measures by the business to ensure that it remains secret

Courts considering whether information is a trade secret, whether an action involves the misuse of a trade secret, and how to compensate an owner of a trade secret for its misuse look at factors including the following:

- The measures taken to maintain secrecy
- The value of the information
- The cost in money or time of creating or developing the information
- The ease with which the information could be acquired or developed by others independently
- The degree to which the owner regards and treats the information as confidential
- The degree to which the recipient regards and treats the information as confidential
- Whether the recipient ought to have known that the information was confidential
- Whether misuse of the information resulted in detriment to the owner

Generally, trade secrets are used to do several things:

1. Ensure an invention or a design is not disclosed to the public before an application for a patent or an industrial design has been made.
2. Protect an invention through means other than patent protection.
3. Protect valuable business information that is not formally protected through other intellectual property (IP) rights.

[99] http://www.ic.gc.ca/eic/siTe/cipointernet-internetopic.nsf/eng/wr03987.html

6.6.1.4 Industrial Design

Industrial designs are the features of a product that appeal to the eye: the contour of a car hood, the pattern of a knitted sweater, the shape of a computer monitor. Distinctive and attractive features like these give products a competitive edge.

An industrial design is the visual features of shape, configuration, pattern or ornament, or any combination of these features, applied to a finished article. For example, the contour of a car hood, the pattern of a knitted sweater, the shape of a computer monitor, the shape of a table or the shape and decoration of a spoon could be industrial designs. An industrial design must have features that appeal to the eye. Registration is intended to protect the product's appearance—not what it is made of, how it is made or how it works. To be eligible for registration, your design must be original.

6.6.1.5 Trademark

A **trademark** is a combination of letters, words, sounds or designs that distinguishes one company's goods or services from those of others in the marketplace. A trademark is unique. It is important to a company because over time, a trademark comes to stand not only for the actual goods and services you sell, but also for your company's reputation and brand. By registering your trademark, you protect it under law from misuse by others, and you gain exclusive rights to use it for a specified term of years.

As this section is focused on AI, we will only be discussing copyright, patents, and trade secrets for the remainder of this section.

6.6.1.6 Copyright: Ownership

With AI now producing outputs that are facially indistinguishable from works of human authorship, it must be established whether copyright subsists or ought to subsist in such AI-generated works or whether they more appropriately belong in the public domain. If it is determined that copyright does or should attach to the outputs of AI systems, it remains to be established who or what should be the first owner in the absence of an identifiable human author.[100]

There is a pressing need for adequate and explicit exceptions for text and data mining to ensure that copyright law does not obstruct or distort the research, development, and operation of AI systems. There is also a lack of clarity around the potential primary and secondary copyright liability of programmers, providers, and users of AI systems in relation to infringing AI processes and outputs.

[100] Martin-Bariteau, Florian; Scassa, Teresa (2021). *Artificial Intelligence and the Law in Canada.* LexisNexis Canada.

Table 6.3 Copyright and Ai—Risks and Opportunities

Risks	Opportunities
• By encouraging the production and exploitation of copyright-protected AI-generated works could exacerbate the risk of human creators being crowded out of the cultural landscape • The field of copyright could end up being cluttered with AI-generated works that human creators must avoid substantially replicating for fear of liability ("copyright landmines") • Copyright overprotection poses a risk to the research, development, training, and transparency of AI technologies • Without adequate limits and exceptions to shield text and data mining from copyright liability, there is a risk that copyright law could stymie technological development and exacerbate problems caused by incomplete, superficial, or biassed training datasets	• Copyright protection could encourage the creation of literary, dramatic, musical, and artistic works, which is the primary purpose of copyright • Without the ongoing investment of human intellectual effort and the limits imposed by copyright law, the capacity to produce AI-generated works could result in a windfall for the public domain • Machine learning and text and data mining technologies offer enormous promise when it comes to creative problem solving and the generation of new knowledge • As we are currently at a moment in time when the technology is developing and norms have become unsettled, we have the opportunity to ensure that a balanced copyright system offers all the necessary incentives and provides the vital space for the ongoing development of AI technologies for the benefit of the public interest

AI and machine learning are playing an increasingly pervasive role in the automated enforcement of copyright online in a manner that threatens to undermine due process and upset the traditional balance of copyright owners' and users' rights. There is an increasingly urgent need to identify and assess the potential impact of sophisticated AI technologies on the pursuit of copyright policy objectives, and to develop legislative and regulatory responses that ensure copyright's substantive technological neutrality as AI continues to evolve.

The rapid evolution of AL technologies presents various risks and opportunities within the copyright field. The implications pertaining to the volume and rapidity with which AI can produce its outputs—and the increasingly elusive distinction between AI-generated and human-authored works—raises issues around the major overarching concern that AI systems could displace human creators as the source of original cultural works. We have sorted some of those risks and opportunities into the tabled list below (Table 6.3).

If there is no change to current copyright strategies, then copyright law may simply be expanded in order to capture the economic value of AI technologies and their outputs. This would likely result in copyright law protecting corporate interests instead of authors, restricting user rights, and impoverishing the public domain overall. With respect to AI and automated copyright enforcement, a critical gap also exists between our carefully crafted copyright law and the increasingly pervasive nature of algorithmic control.

Copyright Law Hypothetical

Facts

A computer uses AI to generate a novel. The novel is wildly popular, selling out in bookstores around the world. As a result of the success and the social impact of the novel, it wins many awards and achievements, and finally becomes a cultural masterpiece.

Questions

1. Should the AI-generated novel be protected by copyright?

If we determine that it should be given copyright protection, then...

2. Who, if anyone, should be deemed to be the author? Or in whom should first ownership of the copyright vest?

If new protections are to be afforded to AI-generated works—as a matter of evidence-based policy-making—then...

3. What would be the appropriate scope and duration for those rights or copyright protections?

If it turns out that the AI-generated novel itself was a creation stemming from copyright infringement carried out by AI systems and beyond the control of the AI developer or user, then...

4. What degree of involvement or control by the programmers, providers, or users of AI could produce direct or indirect liability? That is, at what point is the programmer, provider, or user responsible for the infringing act?

These issues cannot be satisfactorily addressed, however, until we tackle the larger gap in our normative and conceptual thinking about the appropriate interaction of AI and copyright law. Specifically, the extent to which the copyright system can and should play a role in encouraging, facilitating, restricting, and/or regulating the ongoing evolution of artificial intelligence.

6.6.1.7 Patents: Knowledge

Patent law promotes the disclosure of information about inventions and can reward such disclosure with a patent. If AI or AI-generated artefacts are not patentable, for whatever reason, then disclosure of information about them might not occur. One of the issues that arises in patent law relates to whether AI technology can satisfy the already existing statutory requirements for a patent, that is, whether intelligent agents are statutory subject-matter in patent law. If they are, then it must also be determined what the applicable disclosure requirements would be.

Another issue arises when intelligent agents become either an aid to human inventors and decision-makers or become inventors and decision-makers themselves. It remains unclear whether intelligent agents can be "inventors" in patent law. If intelligent agents do not meet the statutory requirements for a patent, then patents will not be issued. If intelligent agents can be deemed to be inventors, then it must be determined whether AI-generated subject matter can be an invention of intelligent agents, or whether the programmers or operators would be the inventors of the products of that agent.

If intelligent agents can invent, then it will be necessary to determine whether subject-matter is patentable if it is obvious to skilled intelligent agents. Adding to the complexity, if intelligent agents do not need patents in order to motivate them to invent, then patents will also not be issued for those products.

6.6.1.8 Trade Secrets: Accountability

The central issues with trade secret law are whether it can protect AI algorithms and the data needed to create machine learning algorithms and, if so, how to ensure the disclosure of meaningful information about how AI algorithms affect individuals. Trade secret law currently protects secret AI algorithms that are used to make decisions from unauthorised disclosure, potentially hindering the accountability of such decision-making.

Greater certainty is required as to the circumstances in which algorithms and training data can be protected as trade secrets. It also remains unclear whether programmers and intelligent agents can be held accountable for the decisions of intelligent agents, as accountability requires that meaningful information be disclosed about the nature and operation of intelligent agents, what they invent, and how they decide.

Some of the main questions in patent law and trade secret law, and relating to AI, are:

1. Are intelligent agents statutory subject matter?
2. What are the disclosure requirements for intelligent agents?
3. Can algorithms and training data be protected as trade secrets?
4. Can intelligent agents invent?
5. Is AI-generated subject matter an invention of (AI) intelligent agents?
6. Is subject-matter patentable if it is obvious to skilled intelligent agents?
7. Are programmers and intelligent agents accountable for the decisions of (AI) intelligent agents?

Patent law provides an incentive to invent, but independent intelligent agents, who may exist within a few decades, will not need such incentives to invent. Resultantly, patents will not likely be granted for their inventions, and disclosure may not occur. Since AI should benefit and be accountable to humans, regulations could be introduced to require intelligent agent inventors to disclose their inventions so as to help create a greater understanding of the inventions and their effects on individuals. As an alternative, patent law could be repurposed to allow for a third party patent on an invention of an intelligent agent in return for the disclosure of the invention.

Trade secrets in AI algorithms may prevent the disclosure of information about automated decision-making, diminishing accountability. To address this, regulations could be introduced to ensure that meaningful information about automated decision-making be provided to individuals. The regulation could also require the submission of information in confidence to a regulatory body to assess the automated decision-making for legal compliance.

6.6.2 Artificial Intelligence and Contract Law

Contracts are an essential regulator of both commercial and social relationships. Our growing dependence on the Internet and the increased digitization of our lives have pushed contracts from a purely commercial realm into virtually every aspect of contemporary life. From work, to education, to filing taxes, to purchasing goods and services, to leisure—most, if not all, of our activities are now governed by standard-form contracts.

The reliance on autonomous agents brings risks to controllers, especially if the autonomous agent goes rogue. Similarly, immutable systems, such as blockchain, present compliance and practical risks to controllers. A legal framework needs to be developed to ensure that the parties using AI—both generally and in the contracting process—cannot inappropriately limit or avoid liability by contract.

Contract law doctrine should be rebalanced to minimise the risk of AI amplifying and exploiting power and information imbalances between contractual parties, further negatively impacting individuals. A transparency and accountability framework for the use of AI in contracting and automated decision-making systems should also be developed, with respect to how they are used in the contracting process, the parameters of the AI actions, and the beneficiaries.

6.6.3 Data Protection, Privacy, and Equality

Artificial intelligence, or AI, drives the demand for large quantities of data, including personal and human behavioural data. The links between personal data and many artificial intelligence applications raise privacy concerns, as well as ethics and human rights issues. Many nations around the world have public and private sector data protection laws which are outdated, with numerous calls for legislative reform. These reforms must address both the need of the organisation to access the large quantities of data required for AI innovation as well as the imperative to properly protect the human right to privacy and data protection.

In the absence of legislative amendment, there is a need to interpret and apply existing data protection laws to the emerging issues raised by the collection, use, and disclosure of data for artificial intelligence purposes. There is also a need to comprehensively review federal and provincial laws that enable or authorise data sharing to ensure that they include safeguards and limitations adapted to the concepts of big data and artificial intelligence. Outdated public and private sector data protection laws must be amended to better adapt to big data and artificial intelligence challenges, and to provide greater clarity.

Policymakers will need to find a way to balance government and business needs for personal data for innovation in artificial intelligence with privacy rights, as well as with related ethical and human rights considerations.

For example, the European Union's *General Data Protection Regulation* (GDPR) has strengthened privacy protections and has provided new rights tailored specifically to the contexts of big data, artificial intelligence, and automated decision-making. The high demand for data and the need for ways to share data that respect both privacy and ethical considerations have also driven a growing interest in new data-sharing frameworks, some of which could be supported by these legislative amendments.

6.6.3.1 Data Protection

The demand for data as the raw material for artificial intelligence creates challenges for data protection and privacy law. For countries in which public and private sector data protection laws have not been substantially updated in decades, those challenges are acute. From a privacy rights perspective, the reform of data protection laws must reinforce the human rights basis for privacy protection and must provide individuals with new rights to better enable them to assert control over their personal data. Although these changes may reinforce consent, they must also address the many ways in which consent has become impractical or a burden for individuals and does not address group harms. Measures are needed to directly address harmful or inappropriate uses of personal data—regardless of consent. As well, data protection laws are a logical place to incorporate protections for the use of personal data in artificial intelligence systems. To be truly meaningful, these reforms must also be accompanied by significant penalties that reinforce individual rights and give privacy commissioners better tools to ensure compliance with privacy obligations.

6.6.3.2 Health Care

There is a strong interest in the use of personal health information in artificial intelligence, including the use of AI to diagnose and treat diseases, analyse images, predict disease outbreaks, or develop personalised medicine. Artificial intelligence could also be used to manage health care system resources and to make decisions about who should be offered which services. In public health care systems, such as the public health system in Canada, there is considerable personal health information in the custody of public sector actors, combined with a high demand for the use of this data.[101]

There is a long and well-established history of sharing de-identified health data with researchers and private sector actors for research and innovation purposes. Such data sharing often relies upon de-identification to protect privacy. Considerable data-sharing infrastructure is therefore already in place at hospitals and in health care agencies. However, issues still arise around the need for consent from new users of existing data and the cross-sectional sharing of data. When massive quantities of de-identified data from multiple sources are combined, the risk of re-identification through AI rises.

In health care, as in other areas in which personal data is used to train AI, there are issues that go beyond privacy. These include concerns over bias in the datasets relied upon, transparency, and accountability. Attention to these complex and intertwined issues may require more than traditional data protection approaches.

6.6.3.3 Equality by Design

Artificial intelligence is increasingly offered by proponents of its use as a tool for overcoming human biases, errors in judgement, and irregularities in decision-making, promising to provide a greater measure of predictability and consistency in decisions and in the rule of law. However, for critics, generating consistency in decision-making is a means of entrenching and amplifying discriminatory practices in a way that perpetuates and reproduces privilege and inequality in human relations through the use of pattern discrimination and predictive analytics while simultaneously rendering these practices less transparent, explainable, and contestable.

Algorithms feed neural networks with neo-liberalism's relations of power and inequality, giving them the capacity to reproduce bias and effectively pass off the descriptive output of "code" as "law", and such "law" as the normative aspirations of justice.

If the rule of law is to retain its comparative advantage, an enforceable "equality by design" default for computer programming—consistent with the aim of the "human-centred" approach to AI governance, is required.[102]

Regulators must find ways to ensure that technological pattern discrimination does not automate and amplify discrimination against individuals and groups through applications of predictive algorithms and artificial intelligence. It is necessary to determine the extent to which mandating value-based design within

---

[101] Martin-Bariteau, Florian; Scassa, Teresa (2021). *Artificial Intelligence and the Law in Canada.* LexisNexis Canada.

[102] "Equality by Design" borrows from the well-established "Privacy by Design"principles and argues for similar strengthened implementation in domestic law and policies for equality as the default value-based inclusive design for machine-learning algorithms and predictive analysis.

regulatory standards can safeguard against automating inequality and injustice in algorithmic decision-making. Given the lack of transparency in data collection and processing, and the limited success of current equality claims, it is important to understand the potential implications of shifting defaults and whether design defaults can be used to better protect and promote equality rights.

In order to meet international legal obligations, to ensure fairness in the algorithmic treatment of all peoples as data subjects, and to advance structural accountability for enforcing substantive equality, legislators should embed an "Equality by Design" model in artificial intelligence regulation.

6.6.4 Human Rights, Violence, and Legal Ethics

The human rights impacts of artificial intelligence are complex. The full range of civil, political, economic, social, and cultural rights protected by international and domestic human rights laws are affected by current uses of this family of technologies. These impacts vary considerably based on how AI is implemented cases.

The human rights impacts of artificial intelligence must be assessed on a case-by-case basis, as a specific algorithmic system's impacts—both positive and negative—depend on factors that are particular to each case. The impacts of a given AI system can vary between rights holders, with some experiencing positive impacts on their enjoyment of a particular right, while others experience negative impacts.

Algorithmic systems can perpetuate bias and discrimination, but the law of discrimination has, for the most part, not yet had to grapple with how algorithmic discrimination can be proven. Public and private sector entities should be required by law to assess the human rights impacts of artificial intelligence prior to the deployment of such systems, and to submit to transparency and oversight measures throughout their operational life.

6.6.4.1 Technology-Facilitated Violence and Abuse

Artificial intelligence is being used, and in some cases specifically designed, to cause harms to members of equality-seeking communities. These equality harms have both individual and collective effects, emanating from both "direct violence" and "structural violence". Technology, such as AI, facilitates violence, re-entrenching social inequalities that undermine the rights of members of equality-seeking communities to self-determination, self-representation, and dignity. Technology-facilitated violence and abuse (TFVA) is most commonly associated with direct violence and individual bad actors, obfuscating the equality harms, which are presented by structural forms of technology-facilitated violence and abuse.[103]

[103] Martin-Bariteau, Florian; Scassa, Teresa (2021). *Artificial Intelligence and the Law in Canada.* LexisNexis Canada.

Existing laws do not adequately capture non-consensual deepfakes, nor are they framed to address the tightly interwoven equality and privacy harms that are a central feature of discriminatory algorithimic profiling by corporations. Meaningful responses to technology-facilitated violence and abuse require reconceptualizing it to include both direct and structural violence, encompassing individual and corporate uses and abuses of artificial intelligence.

6.6.4.2 Legal Ethics and Judicial Decision-Making

Many professional conduct rules that govern lawyers were not designed with legal technology in mind. As a result, these rules are not optimally responsive to the ethical issues that arise when lawyers use AI tools. Current regulatory mechanisms, including those put in place by regional law societies as well as those used in the areas of privacy and consumer protection, do not specifically contemplate public-use legal technologies.

The complexity and dynamism of legal AI tools create challenges for regulation within these societies and regulatory bodies. Because of this, it can be challenging for regulators to determine the right balance between facilitating access to justice and protecting the public from potential harm arising from the use of legal AI tools.

Although, currently, there is not a significant amount of legal AI being used by either lawyers or the public, it is not premature to start thinking very seriously about and taking action on, the ethical issues raised by this technology in the context of legal services. There are already some pockets of meaningful use and these pockets are likely to expand to larger plains in the near future.

Regarding judicial decision-making, it is important to distinguish between when decision-makers are using artificial intelligence and when they are being replaced by it. An appropriate level of reliance should be placed on AI, in order to avoid the tendency to overthrust outcomes provided by AI without weighing them against other evidence, experience, and knowledge.

For these reasons, algorithmic biases should be identified to avoid them leading to inaccurate or discriminatory decisions. As well, different kinds of algorithmic opacity must be understood, as well as how they interact with private interests and the responses required for each of them.

Lawyers around the world are beginning to use statistics, machine learning, and data science to review contracts, investigate case law, or predict judicial outcomes. This ability to mine law as a form of data mining is known as legal analytics. Legal analytics promises to render legal analysis scalable and to complement, rather than substitute, traditional legal research methodologies.

Legal analytics requires an open, but regulated, legal data regime, which provides access to legal information in bulk while also protecting privacy. To ensure consumer confidence, legal analytics must address potential biases and add explanation and transparency to black box algorithms. From description to prediction to prescription, different uses of legal analytics raise different policy concerns and may require different regulatory strategies.

6.7 Social Responsibility

We have previously discussed the concept of social responsibility through the corporate lens; that is, corporate digital social responsibility. This continues to be an ongoing and developing issue, not just for corporations, as governments, agencies, and other organisations work to ensure that their efforts also fulfil socially aware and diverse needs. In this section, we will discuss geopolitical and cultural power dynamics including digital colonialism, digital apartheid, digital redlining and censorship, shutdowns, and blackouts.

6.7.1 Digital Colonialism

The politics of the empire are already permeating the shared histories. Unequal social relations between colonisers and colonised peoples materialised through exploitation, segregation, epistemic violence, and so on. Throughout the world, these discourses of colonialism dominated peoples' perceptions and cultures. Postcolonial critics contended how colonised peoples could attain cultural, economic, and social agency against the oppressive structures and representation imposed on their lives and societies.[104]

However, through temporality, the preface "post" implies that the historical period of colonisation has ended, and the colonised subjects are now free of its discourses. Scholars have focused on the continuity of colonialism even if it has historically ended. The neo-colonial structures and discourses are already part of the different "postcolonial" cultures.[105] The postcolonial era in which colonised countries have gained independence and autonomy has been a means for the populace to regain their own self-determination and freedom. Yet, the neo-colonial structures that are still rampant in the postcolonial societies. Although the nation-state might forward the idea of autonomy and self-determination, new forms of colonialism are always emerging. This dialectic between colonialism and self-determination encompasses a range of fields, changing in form, focus, and scope over time. It is reflected in the complex political and policy relationships between "postcolonial" peoples and the state, especially since most states are replicating the legal and political systems of their former colonisers.

---

[104] Spivak, G. 1999. Can the subaltern speak? In *Social Theory: The Multicultural and Classic Readings* edited by C. Lemert. Boulder, CO: Westview Press. pp. 610–614; Said, E. 1979. *Orientalism*. New York: Vintage Books; Fanon, F. 1963. *The Wretched of the Earth*. New York: Grove Press.

[105] Appiah, K. A. 1991. "Is the Post- in Postmodernism the Post- in Postcolonial?" *Critical Inquiry*, vol 17. pp. 336–357; Shohat, E. 1992. "Notes on the "Post-Colonial." *Social Text*, vol 31. pp. 99–113; Benharrousse, R. 2020 "The Dilapidated Prefix: Beyond Postcolonialism's 'Post' and Towards the Process." In *Post-colonial Praxis: Ramifications and Intricacies*. Mumbai: Notion Press. pp. 36–57.

History articulates that state policy in fields as diverse as health, education, housing, public works, employment, and justice had, and continue to have, negative effects on indigenous peoples after independence.[106] This negative effect is shared throughout the former colonised peoples. Alongside these political tensions, economic interests have manipulated legal and governance frameworks to extract value and resources from former colonised territories, often without adequate compensation or consultation to impacted individuals and communities.[107] Accordingly, Digital Colonialism emerges as a dominant discourse in the digital sphere.

Digital colonialism is a structural form of domination exercised through the centralised ownership and control of the three core pillars of the digital ecosystem: software, hardware, and network connectivity. The control over the latter three gives giant corporations an immense political, economic, and social power over not only individuals, but even nation-states.[108] Assimilation into the tech products, models, and ideologies of foreign powers constitutes a colonisation of the Internet age.[109]

Today, a new form of corporate colonisation is taking place. Instead of the conquest of land, Big Tech corporations are colonising digital technology. The following functions are dominated by a handful of multinational companies: search engines (Google); web browsers (Google Chrome); smartphone and tablet operating systems (Google Android, Apple iOS); desktop and laptop operating systems (Microsoft Windows); office software (Microsoft Office, Google Docs); cloud infrastructure and services (Amazon, Microsoft, Google, IBM); social networking platforms (Facebook, Twitter); transportation (Uber, Lyft); business networking (Microsoft LinkedIn); streaming video (Google YouTube, Netflix, Hulu); and online advertising (Google, Facebook)—among others. These include the five wealthiest corporations in the world, with a combined market cap exceeding $3 trillion.[110] If any nation-state integrates these Big Tech products into their society, these multinational corporations will obtain enormous power over their economy and create technological dependencies that will lead to perpetual resource extraction. This resembles the colonial period in which the colonies were made to be dependent on the coloniser's economy for further exploitation.

Under digital colonialism, digital infrastructure in the Global South is engineered for the Big tech companies' needs, enabling economic and cultural domination while

---

[106] Valaskakis, G.G. 2005. *Indian Country: Essays on Contemporary Native Culture.* Waterloo: Wilfrid Laurier University Press.

[107] Alfred, T. 2009. *Wasáse: Indigenous pathways of action and freedom.* Toronto: University of Toronto Press.

[108] Kwet, M. 2019 "Digital Colonialism: US Empire and the New Imperialism in the Global South," *Race & Class* Volume 60, No. 4 (April). pp. 3–26.

[109] Kwet, M. 2018 "Break the hold of digital colonialism," Mail & Guardian, 29 June. https://mg.co.za/article/2018- 06-29-00-break-the-hold-of-digital-colonialism

[110] The Daily Records. 2018. "Top 10 Largest Companies in the World by Market Cap," 26 March. http://www.thedailyrecords.com/2018-2019-2020-2021/world-famous-top-10-list/highest-selling-brands products-companies-reviews/largest-companies-world-by-market-cap-most-valuable/12829.

imposing privatised forms of governance.[111] To accomplish this task, major corporations design digital technology to ensure their own dominance over critical functions in the tech ecosystem. This allows them to accumulate profits from revenues derived from rent; to exercise control over the flow of information, social activities, and a plethora of other political, social, economic, and military functions which use their technologies.

Digital colonialism depends on code. In *Code: And Other Laws of Cyberspace*, Lawrence Lessig famously argued that computer code shapes the rules, norms, and behaviours of computer-mediated experiences. As a result, "code is law" in the sense that it has the power to usurp legal, institutional, and social norms impacting the political, economic, and cultural domains of society. This critical insight has been applied in fields like copyright, free speech regulation, Internet governance, blockchain, privacy, and even torts.[112] This is similar to architecture in physical space during colonialism. Buildings and infrastructures were built to reinforce the dominance and reach of colonialism.[113]

"Postcolonial" peoples, then, face multiple digital limitations in their access and use of the networked digital infrastructures. The latter threatens to reflect and restructure existing relations of social inequality grounded in colonialism and continuing processes of neo-colonialism. Indigenous peoples are acutely aware of this potential, and so are working with various partners to decolonize the digital sphere. They are undertaking a variety of projects that represent their diverse and localised experiences, alongside a common desire for self-determination.

Rural and remote indigenous communities face persistent access problems to the digital associated with the historic and ongoing effects of colonialism. Remote indigenous communities are becoming "offline by design" because their going online has been challenged.[114] Indigenous peoples are asserting their digital self-determination by using these platforms to build online communities, express virtual identities, and represent their culture virtually. Hence, they are no longer static as offline, but becoming "networked individualism".[115] Their engagement with the digital sphere resists the imposed representations of their identities and deterritorialized conceptions of virtual communities. Accordingly, the former colonised peoples are always engaged in the process of decolonizing the latent neo–/colonial discourses which are dominating the Internet.

---

[111]Bakan, J. 2005. *The Corporation: The Pathological Pursuit of Profit and Power*. New York, NY: Free Press.

[112]Lessig, L. 1999. *Code and Other Laws of Cyberspace*. New York: Basic Books.

[113]Reidenberg, J. 1998. "Lex Informatica: The Formation of Information Policy Rules Through Technology," Texas Law Review 76, no. 3.

[114]Sandvig, C. 2006. *The Structural Problems of the Internet for Cultural Policy. In Critical Cyberculture Studies* edited by D. Silver and A. Massanari. New York: NYU Press. pp.107–118.

[115]Raine, L. & Wellman, B. 2012. *Networked: The New Social Operating System*. Cambridge and London: The MIT Press.

6.7.2 Digital Apartheid

Digital apartheid has also been a key concept in debates around digital self-determination. For authors such as Christian Fuchs, digital apartheid means that "certain groups and regions of the world are systematically excluded from cyber-space and the benefits that it can create."[116]

Brown and Czerniewicz (2010), drawing on a research project interrogating the access of higher education students in South Africa to information and communications technology (ICT), highlight that while age or generational aspects have been a characteristic of digital divides, now the latter are rather a question of access and opportunity, claiming that in the present day, "digital apartheid is alive and well."[117]

Borrowing from Graham (2011),[118] and extending to the representation of conditions surrounding higher education in post-apartheid South Africa, Ashton et al. (2018)[119] highlight the concept of digital apartheid as a multidimensional process with three dimensions—a material dimension (including access to infrastructure, devices, cellular coverage, electricity), a skills dimension (including education legacy regarding computer training, social capital with regard to the family/community computer skills), and a virtual dimension (including language, culture, and contextual relevance). The authors argue that "The virtual dimension emerges from the intentional act of "digital redlining", which takes on a number of forms. It may be under the guise of protecting an organisation from spam and illicit, harmful cyber-attacks, but has the secondary outcome of blocking or filtering out communities who only have access through cheaper portals."[120] It also includes the influence of the Westernised, English Internet, that further influences content visibility.

The skills dimension emerges from an understanding that ICT lessons were not a part of the curriculum until recently and therefore the skill development remained underexposed and restricted. The authors refer to the material dimension as the most cited concern regarding introducing technology as part of the curriculum, arguing that "the lack of power infrastructure in lower socio-economic areas and

---

[116]Finnemann, Niels Ole (2010). "Christian Fuchs: Internet and society—Social theory in the information age. London: Routledge. 2008". *MedieKultur: Journal of Media and Communication Research*. 26 (48): 4.

[117]Brown, C.; Czerniewicz, L. (2010). "Debunking the 'digital native': beyond digital apartheid, towards digital democracy". *Journal of Computer Assisted Learning*. 26 (5): 357–369.

[118]Graham, Mark (2011). "Time machines and virtual portals". *Progress in Development Studies*. 11 (3): 211–227.

[119]Barnard-Ashton, Paula; Adams, Fasloen; Rothberg, Alan; McInerney, Patricia (2018). "Digital apartheid and the effect of mobile technology during rural fieldwork". *South African Journal of Occupational Therapy*. 48 (2): 20–25.

[120]Barnard-Ashton, Paula; Adams, Fasloen; Rothberg, Alan; McInerney, Patricia (2018). "Digital apartheid and the effect of mobile technology during rural fieldwork". *South African Journal of Occupational Therapy*. 48 (2): 20–25.

exorbitant data costs, impact some students' ability to access their learning resources."[121]

6.7.2.1 Example: Digital Apartheid in India

Since 2019, this concept—signifying advantages to some and dispossession of some others—has also been used to characterise Internet shutdowns and communications blockades in Jammu and Kashmir.[122] The region, contested and claimed by both India and Pakistan in its entirety and a site of an active armed conflict, witnessed the Indian State imposing a total communication blackout and Internet shutdown in Jammu and Kashmir on the intervening nights of 4 and 5 August 2019 as part of its unilateral measures to remove the semi-autonomous nature of the disputed territory of Jammu and Kashmir. Low speed 2G Internet was restored in January 2020[123] while high speed 4G Internet was restored in February 2021.[124] A 2019 report notes that between 2012 and 2019, there have been 180 Internet shutdowns in the region.[125] India also topped the list of 29 countries that had disrupted access to the Internet for their people in the year 2020.[126] The report by Access Now highlighted, "India had instituted what had become a perpetual, punitive shutdown in Jammu and Kashmir beginning in August 2019. Residents in these states had previously experienced frequent periodic shutdowns, and in 2020 they were deprived of reliable, secure, open, and accessible Internet on an ongoing basis."[127]

In placing these frequent shutdowns in the context of the ongoing conflict in Kashmir, the report *Kashmir's Internet Siege* (2020) by the Jammu Kashmir Coalition of Civil Society argues that with the frequent Internet shutdowns, the Indian government has been enacting in these regions a "digital apartheid," "a form of systemic and pervasive discriminatory treatment and collective punishment." According to the report, "frequent and prolonged Internet shutdowns enact a profound digital apartheid by systematically and structurally depriving the people of Kashmir of the means to participate in a highly networked and digitised world."[128]

---

[121] Barnard-Ashton, Paula; Adams, Fasloen; Rothberg, Alan; McInerney, Patricia (2018). "Digital apartheid and the effect of mobile technology during rural fieldwork". *South African Journal of Occupational Therapy*. 48 (2): 20–25.

[122] "Article 370: What happened with Kashmir and why it matters". *BBC News*. 2019-08-05.

[123] "2G mobile Internet services restored in J&K". *The Hindu*. Special Correspondent. 2020-01-25.

[124] Jain, Bharti; P, M. Saleem (February 6, 2021). "After 18 months, 4G internet services restored in J&K | India News - Times of India". *The Times of India*.

[125] Maqbool, Majid (2019-10-22). "Internet shut down 180 times in J&K over the past 8 years". *National Herald*.

[126] Chakravarti, Ankita (March 4, 2021). "India saw the highest number of internet shutdowns in the world in 2020". *India Today*.

[127] Chakravarti, Ankita (March 4, 2021). "India saw the highest number of internet shutdowns in the world in 2020". *India Today*.

[128] "Kashmir's Internet Siege - an ongoing assault on digital rights". *jkccs.net*. 2020.

This systematic censorship and deprivation not only resulted in excluding the people, collectively, from participating in cyberspace, but as was evident, it crippled IT companies and startups in Kashmir.[129] It was noted to have affected at least a thousand employees working in this sector just in the third month of the world's longest Internet shutdown that began on the intervening nights of 4 and 5 August 2019 across Jammu and Kashmir.[130] In a statement, UN Special Rapporteurs referred to the communication blackout as a collective punishment without any pretext of precipitating an offence. "The shutdown of the Internet and telecommunication networks, without justification from the government, are inconsistent with the fundamental norms of necessity and proportionality," the experts said.[131] A news report quoting the story of an entrepreneur who had been doing well with a startup noted that the "Internet is the oxygen for start-ups. The Centre pulled that plug on August 5. The virtual world was our space for growth. Now that's gone. All employees and producers have been rendered jobless [..] I have to work by hook or by crook to meet the damage inflicted by loss of customers, undelivered orders, and accumulated goods after the non-availability of Internet."[132]

In June 2020, it was reported for the first time how non-local companies were able to bag most contracts online for the mining of mineral blocks, as locals were left at a disadvantage due to the ban on high-speed Internet.[133]

The effect of this digital apartheid was also witnessed during the lockdown induced by the COVID-19 pandemic, leaving the health care infrastructure crippled as doctors complained about not being able to access information or attend trainings on coronavirus owing to the restricted Internet. The president of the Doctors Association noted that the awareness drives that were carried out elsewhere about the virus were impossible to run in Kashmir. "We want to educate people through videos, which is not possible at 2G speed. We are handicapped in the absence of high-speed Internet."[134] Health experts and the locals warned that the Internet blackout was hampering the fight against coronavirus in the region. The Internet shutdown also affected education across all levels in the region. News reports noted how Kashmiri education was left behind even as life elsewhere was moving online in dealing with the stay-at-home guidelines during the pandemic.[135] A news report after a year of the communication blackout and subsequent restriction on high-speed

[129] Maqbool, Majid (2019-10-14). "Kashmir communication shutdown cripples IT companies and start-ups in Valley". *National Herald.*

[130] "Kashmir's Internet Siege - an ongoing assault on digital rights". *jkccs.net*. 2020.

[131] "Kashmir communications shutdown a 'collective punishment' that must be reversed, say UN experts". *UN News*. 2019-08-22.

[132] Ashiq, Peerzada (2019-12-07). "In a land without Internet: How the communication blackout is forcing young entrepreneurs out of Kashmir Valley". *The Hindu.*

[133] "Kashmir: Online Bidding for Mineral Blocks Leaves Locals at a Disadvantage". *The Wire.*

[134] "Lack of internet in Kashmir prevents doctors from fighting coronavirus: Report". *HT Tech.* 2020-03-31.

[135] "As Life Moves Online Amid the Pandemic, Kashmiri Education Is Being Left Behind". *thediplomat.com*

Internet highlighted that it had "ravaged health, education, entrepreneurship" in the region.[136]

6.7.3 Digital Redlining

Digital redlining is the practice of creating and perpetuating inequities between already marginalised groups specifically through the use of digital technologies, digital content, and the Internet.[137] The concept of digital redlining is an extension of the practice of redlining in housing discrimination,[138] a historical legal practice in the United States and Canada dating back to the 1930s[139] where red lines were drawn on maps to indicate poor and primarily black neighbourhoods that were deemed unsuitable for loans or further development, which created great economic disparities between neighbourhoods.[140,141]

The term "redlining" was popularised by Dr. Chris Gilliard—a privacy scholar—who defines digital redlining as "the creation and maintenance of tech practices, policies, pedagogies, and investment decisions that enforce class boundaries and discriminate against specific groups".[142]

Though digital redlining is related to the digital divide and techniques such as web lining and personalization, it is distinct from these concepts as they are part of larger complex systemic issues.[143] It can refer to practices that create inequalities in access to technology services in geographical areas, such as when Internet service providers decide to not service specific geographic areas because they are perceived to be not as profitable and thus reduce access to crucial services and civic participation.[144] It can also be used to refer to inequities caused by the policies and practices

[136] Zargar, Safwat. "A year without high-speed internet ravaged health, education, entrepreneurship in Kashmir". *Scroll.in*.

[137] Gilliard, Christopher (2019-11-21). Prepared testimony and statement for the record of Christopher Gilliard PhD, Hearing on Banking on your Data the Role of Big Data in Your Financial Services, Before the House Financial Services Committee Task Force on Financial Technology. https://www.congress.gov/116/meeting/house/110251/witnesses/HHRG-116-BA00-Wstate-GillardC-20191121.pdf

[138] Podesta, John; Penny, Pritzker; Moniz, Earnest, J.; Holdren, John; Zients, Jeffery (May 2014). "Big Data: Seizing Opportunities, Preserving Values" *Whitehouse.gov*.

[139] Gilliard, Chris (2017-07-03). "Pedagogy and the Logic of Platforms". *er.educause.edu*.

[140] Madrigal, Alexis C. (2014-05-22). "The Racist Housing Policy That Made Your Neighborhood". *The Atlantic*.

[141] Harris, Richard; Forrester, Doris (2003-12-01). "The Suburban Origins of Redlining: A Canadian Case Study, 1935–54". *Urban Studies*. **40** (13): 2661–2686.

[142] Oremus, Will (September 17, 2021). "A Detroit community college professor is fighting Silicon Valley's surveillance machine. People are listening". *Washington Post*.

[143] Bidgoli, Hossein (2004-04-26). *The Internet Encyclopedia, Volume 3 (P - Z)*. John Wiley & Sons.

[144] Neidig, Harper (2017-09-25). "AT&T hit with second complaint of discrimination against low-income neighborhoods". *TheHill*.

of digital technologies.[145] For instance, with these methods, inequities are accomplished through divisions that are created via algorithms which are hidden from the technology user; the use of big data and analytics allows for a much more nuanced form of discrimination that can target specific vulnerable populations.[146] These algorithmic means are enabled through the use of unregulated data technologies that apply a score to individuals that statistically categorise personality traits or tendencies which are similar to a credit score but are proprietary to the technology companies and not under outside oversight.[147]

It is important to note the differences between "digital redling" and "weblining" as well as "digital redlining" and the "digital divide".[148] Weblining refers to the process of charging customers different prices based on profile information while digital redlining focuses not on pricing, but access.[149] As early as 2002 the *Gale Encyclopaedia of E-Commerce* puts forth the distinction more in use today: weblining is the pervasive and generally accepted (or at least tolerated) practice of personalising access to products and services in ways invisible to the user; digital redlining is when such personalised, data-driven schemes perpetuate traditional advantages of privileged demographics.[150] The digital divide refers to one of a number of impacts of digital redlining, and digital redlining is one of a number of ways in which the divide is maintained or extended.[151,152]

6.7.4 Censorship, Shutdowns, and Blackouts

Censorship, information regulation, and isolation from the outside world are the basic components of a government's descent into digital authoritarianism. Indeed, an Internet shutdown is a proven all-in-one tool that achieves ultimate control in a single, swift action.

In 2021, Access Now and the #KeepItOn coalition documented 182 Internet shutdowns across 34 countries. This shows a dramatic resurgence of this oppressive

---

[145] Stepanek, Marcia (2 April 2000). "Weblining". *Bloomberg BusinessWeek.*

[146] O'Neil, Cathy (2016-09-06). *Weapons of Math Destruction: How Big Data Increases Inequality and Threatens Democracy.* Crown/Archetype.

[147] Taylor, Astra; Sadowski, Jathan. "How Companies Turn Your Facebook Activity Into a Credit Score | The Nation". *The Nation.*

[148] Gilliard, Chris; Culik, Hugh (2016-05-24). "Digital Redlining, Access, and Privacy". *Common Sense Education.*

[149] Bidgoli, Hossein (2004-04-26). *The Internet Encyclopedia, Volume 3 (P - Z).* John Wiley & Sons.

[150] Malonis, Jane A. (2002). *Gale Encyclopedia of E-Commerce.* Gale Group. pp. 758–760.

[151] Prieger, James E. (December 2001). The Supply Side of the Digital Divide: Is There Redlining in the Broadband Internet Access Market? *AEI-Brookings Joint Center Working Paper No. 01–16.*

[152] Gilliard, Chris; Culik, Hugh (2016-05-24). "Digital Redlining, Access, and Privacy". *Common Sense Education.*

Table 6.4 Number of Internet shutdowns by region in 2021

Region	Number of shutdowns
Africa	19
Asia Pacific	129
Eastern Europe and Central Asia	7
Latin America and the Caribbean	4
Middle East and North Africa	23

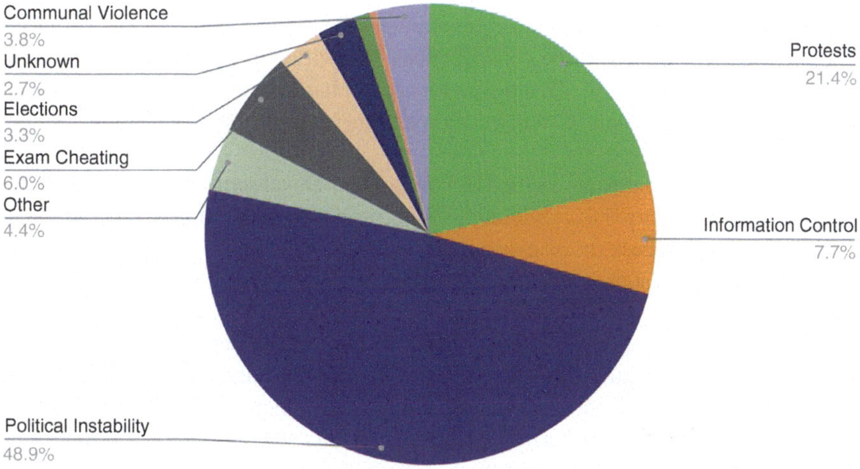

Fig. 6.1 Causes of Internet shutdowns in 2021

form of control compared to 159 shutdowns recorded in 29 countries in 2020.[153] A close look at the 2021 data shows that what could have been a story about the global decline of Internet shutdowns after the pandemic instead reveals a return to the rights-harming tactics of the pre-pandemic period. There was an increase of 23 shutdowns from 2020 to 2021, signalling a continuation of this concerning trend.[154] We can break down these numbers for 2021 by region, as well. This is provided in the following Table 6.4.

The available data on the determined causes of Internet shutdowns, around the world, for 2021, are broken down in the pie chart below (Fig. 6.1). This chart is based on data provided by Access Now in their report, *The Return of Digital Authoritarianism: Internet Shutdowns in 2021.*

[153] The #KeepItOn campaign unites and organises global organisations and efforts to end internet shutdowns. The campaign was launched by a coalition of about 70 organisations in 2016 at RightsCon in Silicon Valley. Membership of the coalition has since increased rapidly to 282 organisations from 105 countries around the world ranging from civil society, rights, and advocacy groups to research centres, detection networks, foundations, and media organisations.

[154] Access Now. *The return of digital authoritarianism: internet shutdowns in 2021.* https://www.accessnow.org/keepiton-2021-report

As we can see in the above figure, the two largest causes of Internet shutdowns, in countries around the world in 2021, were related to political instability and protests.[155]

6.8 Relevance to our Collective Future

Internet shutdowns happen because governments want to control the flow of information online, regardless of the consequences for their own people, economies, and reputation around the world. That desire for control has translated to increased use of network disruptions, even as we see momentum grow in the global fight to stop shutdowns.

In May 2021, participants at the G7 Foreign and Development Ministers' Meeting issued a communiqué to condemn "actions by states to intentionally disrupt their own populations' access to, or dissemination of, information, knowledge, and data online."[156] Heads of state echoed this sentiment in their own *Carbis Bay G7 Summit Communique*, stating, "We also affirm our opposition to measures which may undermine these democratic values, such as government-imposed Internet shutdowns and network restrictions."[157] Notably, this communiqué adopted the definition of an Internet shutdown Access Now and many members of the #KeepItOn coalition use, reflecting familiarity with our campaign and goals, and our presence in events leading to the 2021 G7 Summit.

When the UN Human Rights Council adopted a new version of the resolution on "the promotion, protection and enjoyment of human rights on the Internet" (A/HRC/47/16) in July 2021, it specifically called out shutdowns and mandated a report on "the trend in Internet shutdowns, analysing their causes, their legal implications and their impact on a range of human rights, including economic, social, and cultural rights."[158]

---

[155] Access Now. "Internet Shutdowns in 2021" in *The Return of Digital Authoritarianism*. From: https://www.accessnow.org/cms/assets/uploads/2022/05/2021-KIO-Report-May-24-2022.pdf

[156] Access Now (2021). G7 leaders decry internet shutdowns, but leave personal data up for grabs. From https://www.accessnow.org/g7-internet-shutdowns-personal-data/

[157] G7UK.org (2021). Carbis Bay G7 Summit Communiqué. From https://www.g7uk.org/wp-content/uploads/2021/06/Carbis-Bay-G7-Summit-Communique-PDF-430 KB-25-pages-5.pdf.

[158] United Nations (2021). The promotion, protection and enjoyment of human rights on the Internet: resolution. From https://digitallibrary.un.org/record/3937534

6.9 Summary

In this chapter, we described some of the major emerging topics in data sovereignty and digital governance, including the concept of digital rights, indigenous data sovereignty, the issues associated with data centres and data mines, digital self-determination, artificial intelligence, and social responsibility, including digital colonialism, redlining, and Internet shutdown and censorship.

The answers to the following questions are provided within this chapter:

1. What are some of the causes of the Internet shutdowns that we have seen in 2021?
2. Why might a government use censorship of the Internet during protests?
3. Why are government Internet shutdowns an issue from a human rights perspective?
4. What are some examples of the Internet shutdowns from 2021?
5. What governance strategies are used by the countries which tend to see more occurrences of Internet shutdowns than others?

Additional References

- Clark, Jack (2015b). "Why 2015 Was a Breakthrough Year in Artificial Intelligence". Bloomberg.com. Available online at: https://www.bloomberg.com/news/articles/2015-12-08/why-2015-was-a-breakthrough-year-in-artificial-intelligence
- Crevier, Daniel (1993). AI: The Tumultuous Search for Artificial Intelligence. New York, NY: BasicBooks. pp. 161–162, 197–203, 211, 240.
- Howe, J. (November 1994). "Artificial Intelligence at Edinburgh University: a Perspective".
- Legg, Shane; Hutter, Marcus (15 June 2007). "A Collection of Definitions of Intelligence".
- Martin-Bariteau, Florian; Scassa, Teresa (2021). *Artificial Intelligence and the Law in Canada.* LexisNexis Canada.
- McCorduck, Pamela (2004), Machines Who Think (second ed.), Natick, MA: A. K. Peters, Ltd., pp. 426–441.
- NRC (United States National Research Council) (1999). "Developments in Artificial Intelligence". Funding a Revolution: Government Support for Computing Research. National Academy Press.
- Newquist, HP (1994). The Brain Makers: Genius, Ego, And Greed In The Quest For Machines That Think. New York: Macmillan/SAMS.
- Nilsson, Nils (1998). Artificial Intelligence: A New Synthesis. Morgan Kaufmann.
- Poole, David; Mackworth, Alan; Goebel, Randy (1998). Computational Intelligence: A Logical Approach. New York: Oxford University Press.
- Russell, Stuart J.; Norvig, Peter (2003), Artificial Intelligence: A Modern Approach (second ed.), Upper Saddle River, New Jersey: Prentice Hall.

Conclusion

The twin concepts of data sovereignty and digital governance are shaking up our legal systems, domestically as well as around the world, as nations try to come to terms with a rapidly changing online dynamic.

Cloud computing and cloud service providers have brought about a significant shift in the globalised delivery of information technology services and computing resources. When data is stored in a cloud environment, regardless of where the cloud service resources are physically located, the data may be subject to the laws of other countries. In law, data sovereignty refers to the control a nation—or a state—has over access to and disclosure of its digital information, subject only to its laws. Where a cloud service provider—as a large-scale corporate entity—is operating on a global level, it could end up being required to comply with a court order, warrant, or subpoena request from a foreign law enforcement agency seeking to obtain data relating to the affairs of another nation.

When a crime takes place or a dispute occurs between two parties who reside within the same geographic location, the determination of which law applies to that scenario is easily distinguished as being the law of that geographic location. When a dispute occurs between two parties who reside in different nations—or legal jurisdictions—the methodology for determining which country's laws and legal forum should apply is rooted in the topics of private international law, international criminal law, and conflicts of law. While these topics are by no means straightforward or simple, the strategy for addressing legal disputes or criminal offences which occur within or across legal jurisdictions becomes greatly complexified when the alleged crime, dispute, or other offence takes place online or is otherwise enabled via the Internet.

While jurisdiction is often linked to sovereignty over a territorial location, jurisdiction can also exist without a connection to territory. The type of jurisdictional authority held by a governing body indicates whether that nation or state can undertake enforcement action to uphold its law. This becomes more complex when we consider online activities, particularly cybercrimes, defamation, and

M. Lukings, A. Habibi Lashkari, *Understanding Cybersecurity Law in Data Sovereignty and Digital Governance*, Progress in IS,
https://doi.org/10.1007/978-3-031-14264-2

other tortious activities, as there may not necessarily be a defined geographic area to distinguish which governing authority has the legal jurisdiction to address these matters.

Digital corporate social responsibility extends the social obligations related to the digital governance of a corporation or other governing body to that which is connected to their position within the digital sphere. The legal issues which intersect with and relate to user-generated content can be wide-ranging, including determination of ownership and copyright, personal privacy protection, individual freedom of expression, issues of cyberbullying, illegal content distribution, and large-scale corporate data breaches.

From the context of data sovereignty and digital governance, from a worldwide perspective, it is becoming increasingly necessary to work on creating unified strategies for protecting and preserving private and personal data while also allowing for necessary data to be moved between jurisdictions to allow for cooperative efforts between nations to enable greater progression for us as human beings. Rather than becoming more divisive in our regulations, it is ideal to imagine a world in which data can be transferred and moved safely, without fear of breach or attack. While we are certainly not at a point in which we can look to a grand unification of data protection legislation across all nations, it is absolutely a goal to which we can all aspire.

References

1. Part I of the Constitution Act, 1982. Canadian Charter of Rights and Freedoms, s 7, Part 1 of the Constitution Act, 1982, being Schedule B to the Canada Act 1982 (UK), 1982, c 11.
2. Canadian Human Rights Act (RSC, 1985, c H-6).
3. United Nations. Universal Declaration of Human Rights. 1948.
4. Privacy Act (RSC, 1985, c P-21).
5. Access to Information Act (RSC, 1985, c A-1).
6. Personal Information Protection and Electronic Documents Act (SC 2000, c 5). ("PIPEDA")
7. Protecting Canadians from Online Crime Act (SC 2014, c 31).
8. Copyright Act (RSC 1985, c C-42).
9. Canada's Anti-Spam Legislation (SC 2010, c 23)
10. Criminal Code (RSC (1985), c C-46)
11. Bill C-10
12. Bartsch, R., 2020. The relationship of drug and human trafficking and their facilitation via cyryptomarkets and the dark web: A recommendation for cryptocurrency regulation.
13. The Palgrave Handbook of International Cybercrime and Cyberdeviance
14. Kaur, G., 2021. Understanding Cybersecurity Management in Fintech.
15. Martin-Bariteau, Florian; Scassa, Teresa (2021). *Artificial Intelligence and the Law in Canada.* LexisNexis Canada.
16. Geist, M., 2015. Law, privacy and Surveillance in Canada in the Post-Snowden Era.
17. World Health Organisation Framework Convention on Tobacco Control (United Nations, *Treaty Series*, vol. 2302, No. 41032)
18. Sendai Framework for Disaster Risk Reduction 2015–2030 (resolution 69/283, annex II)
19. United Nations Convention on the Law of the Sea (United Nations, Treaty Series, vol. 1833, No. 31363)

M. Lukings, A. Habibi Lashkari, *Understanding Cybersecurity Law in Data Sovereignty and Digital Governance*, Progress in IS,
https://doi.org/10.1007/978-3-031-14264-2

The manufacturer's authorised representative in the EU is Springer Nature Customer Service Centre GmbH, Europaplatz 3, 69115 Heidelberg, Germany. If you have any concerns regarding our products, please contact ProductSafety@springernature.com

Printed and bound by CPI Group (UK) Ltd, Croydon, CR0 4YY
30/09/2026
02225493-0008